SOIL MECHANICS

A one-dimensional introduction

This introductory course on soil mechanics presents the key concepts of stress, stiffness, seepage, consolidation, and strength within a one-dimensional framework. Consideration of the mechanical behaviour of soils requires us to consider density alongside stresses, thus permitting the unification of deformation and strength characteristics. Soils are described in a way which can be integrated with concurrent teaching of the properties of other engineering materials. The book includes a model of the shearing of soil and some examples of soil-structure interaction which are capable of theoretical analysis using one-dimensional governing equations. The text contains many worked examples, and exercises are given for private study at the end of all chapters. Some suggestions for laboratory demonstrations that could accompany such an introductory course are sprinkled through the book.

David Muir Wood has taught soil mechanics and geotechnical engineering at the universities of Cambridge, Glasgow and Bristol since 1975 and has contributed to courses on soil mechanics in many countries around the world. He is the author of numerous research papers and book chapters. His previous books include *Soil behaviour and critical state soil mechanics* (1990) and *Geotechnical modelling* (2004). He was co-chairman of the United Kingdom GeotechniCAL computer-aided learning project.

Soil mechanics

A ONE-DIMENSIONAL INTRODUCTION

David Muir Wood
University of Bristol

CAMBRIDGE UNIVERSITY PRESS
Cambridge, New York, Melbourne, Madrid, Cape Town, Singapore,
São Paulo, Delhi, Dubai, Tokyo

Cambridge University Press
32 Avenue of the Americas, New York, NY 10013-2473, USA

www.cambridge.org
Information on this title: www.cambridge.org/9780521741323

© David Muir Wood 2009

This publication is in copyright. Subject to statutory exception
and to the provisions of relevant collective licensing agreements,
no reproduction of any part may take place without the written
permission of Cambridge University Press.

First published 2009

Printed in the United States of America

A catalog record for this publication is available from the British Library.

Library of Congress Cataloging in Publication data

Muir Wood, David, 1949–
Soil mechanics : a one-dimensional introduction / David Muir Wood.
    p.   cm.
Includes bibliographical references and index.
ISBN 978-0-521-51773-7 (hardback) – ISBN 978-0-521-74132-3 (pbk.)
1. Soil mechanics. I. Title.
TA710.W5983   2009
624.1′5136 – dc22   2009020445

ISBN 978-0-521-51773-7 Hardback
ISBN 978-0-521-74132-3 Paperback

Cambridge University Press has no responsibility for the persistence or
accuracy of URLs for external or third-party Internet Web sites referred to in
this publication and does not guarantee that any content on such Web sites is,
or will remain, accurate or appropriate.

Contents

Preface	*page* ix

1 **Introduction** 1

1.1	Introduction	1
1.2	Soil mechanics	2
1.3	Range of problems/applications	2
1.4	Scope of this book	10
1.5	Mind maps	11

2 **Stress in soils** 12

2.1	Introduction	12
2.2	Equilibrium	12
2.3	Gravity	13
2.4	Stress	16
2.5	Exercises: Stress	18
2.6	Vertical stress profile	19
	2.6.1 Worked examples	21
2.7	Water in the ground: Introduction to hydrostatics	23
	2.7.1 Worked example: Archimedes uplift on spherical object	26
2.8	Total and effective stresses	28
	2.8.1 Worked examples	32
2.9	Summary	37
2.10	Exercises: Profiles of total stress, effective stress, pore pressure	37

3 **Density** 40

3.1	Introduction	40
3.2	Units	40
3.3	Descriptions of packing and density	41
	3.3.1 Volumetric ratios	43
	3.3.2 Water content	44
	3.3.3 Densities	44

v

	3.3.4 Unit weights	46
	3.3.5 Typical values	46
3.4	Measurement of packing	47
	3.4.1 Compaction	50
3.5	Soil particles	52
3.6	Laboratory exercise: particle size distribution and other classification tests	56
	3.6.1 Sieving	56
	3.6.2 Sedimentation	57
	3.6.3 Particle shape	61
	3.6.4 Sand: relative density	61
3.7	Summary	62
3.8	Exercises: Density	64
	3.8.1 Multiple choice questions	64
	3.8.2 Calculation exercises	65

4 Stiffness 67

4.1	Introduction	67
4.2	Linear elasticity	67
4.3	Natural and true strain	70
4.4	One-dimensional testing of soils	70
	4.4.1 Hooke's Law: confined one-dimensional stiffness ♣	72
4.5	One-dimensional (confined) stiffness of soils	74
4.6	Calculation of strains	78
	4.6.1 Worked examples: Calculation of settlement	79
4.7	Overconsolidation	82
	4.7.1 Worked examples: Overconsolidation	84
4.8	Summary	87
4.9	Exercises: Stiffness	87

5 Seepage 90

5.1	Introduction	90
5.2	Total head: Bernoulli's equation	90
5.3	Poiseuille's equation	96
5.4	Permeability	99
	5.4.1 Darcy or Forchheimer?	102
5.5	Measurement of permeability	104
5.6	Permeability of layered soil	106
5.7	Seepage forces	108
5.8	Radial flow to vertical drain	111
5.9	Radial flow to point drain	112
5.10	Worked examples: Seepage	113
	5.10.1 Example: flow through soil column	113
	5.10.2 Example: effect of changing reference datum	116

Contents

    5.10.3 Example: pumping from aquifer   117
    5.10.4 Example: flow into excavation   119
  5.11 Summary   121
  5.12 Exercises: Seepage   123

6 Change in stress 127
  6.1 Introduction   127
  6.2 Stress change and soil permeability   127
  6.3 Worked examples   130
    6.3.1 Example 1   130
    6.3.2 Example 2   131
    6.3.3 Example 3   133
  6.4 Summary   134
  6.5 Exercises: Change in stress   136

7 Consolidation 138
  7.1 Introduction   138
  7.2 Describing the problem   140
  7.3 Parabolic isochrones   142
  7.4 Worked examples   149
    7.4.1 Example 1: Determination of coefficient of consolidation   149
    7.4.2 Example 2   152
    7.4.3 Example 3   154
    7.4.4 Example 4   155
  7.5 Consolidation: exact analysis ♣   155
    7.5.1 Semi-infinite layer   159
    7.5.2 Finite layer   161
  7.6 Summary   165
  7.7 Exercises: Consolidation   167

8 Strength 169
  8.1 Introduction   169
  8.2 Failure mechanisms   169
  8.3 Shear box and strength of soils   171
  8.4 Strength model   173
  8.5 Dilatancy   174
  8.6 Drained and undrained strength   177
  8.7 Clay: overconsolidation and undrained strength   179
  8.8 Pile load capacity   181
  8.9 Infinite slope   185
    8.9.1 Laboratory exercise: Angle of repose   191
  8.10 Undrained strength of clay: fall-cone test   193
  8.11 Simple model of shearing ♣   195
    8.11.1 Stiffness   196

8.11.2 Strength	197
8.11.3 Mobilisation of strength	197
8.11.4 Dilatancy	198
8.11.5 Complete stress:strain relationship	199
8.11.6 Drained and undrained response	200
8.11.7 Model: summary	203
8.12 Summary	203
8.13 Exercises: Strength	205

9 Soil-structure interaction ... 208

9.1 Introduction	208
9.2 Pile under axial loading ♣	211
9.2.1 Examples	215
9.3 Bending of an elastic beam ♣	216
9.4 Elastic beam on elastic foundation ♣	220
9.5 Pile under lateral loading ♣	224
9.6 Soil-structure interaction: next steps	226
9.7 Summary	227
9.8 Exercises: Soil-structure interaction	227

10 Envoi ... 230

10.1 Summary	230
10.2 Beyond the single dimension	231

Exercises: numerical answers ... 232

Index   237

Preface

This book has emerged from a number of stimuli.

There is a view that soils are special: that their characteristics are so extraordinary that they can only be understood by a small band of specialists. Obviously, soils do have some special properties: the central importance of density and change of density merits particular attention. However, in the context of teaching principles of soil mechanics to undergraduates in the early years of their civil engineering degree programmes, I believe that there is advantage to be gained in trying to integrate this teaching with other teaching of properties of engineering materials to which the students are being exposed at the same time.

It is a fundamental tenet of *critical state soil mechanics* – with which I grew up in my undergraduate days – that consideration of the mechanical behaviour of soils requires us to consider density alongside effective stresses, thus permitting the unification of deformation and strength characteristics. This can be seen as a broad interpretation of the phrase *critical state soil mechanics*. I believe that such a unification can aid the teaching and understanding of soil mechanics.

There is an elegant book by A. J. Roberts[1] which demonstrates in a unified way how a common mathematical framework can be applied to problems of solid mechanics, fluid mechanics, traffic flow and so on. While I cannot hope to emulate this elegance, the title prompted me to explore a similar one-dimensional theme for the presentation of many of the key concepts of soil mechanics: density, stress, stiffness, strength and fluid flow.

This one-dimensional approach to soil mechanics has formed the basis for an introductory course of ten one-hour lectures with ten one-hour problem classes and one three-hour laboratory afternoon for first-year civil engineering undergraduates at Bristol University. The material of that course is contained in this book. I have added a chapter on the analysis of one-dimensional consolidation, which fits neatly with the theme of the book. I have also included a model of the shearing of soil and some examples of soil-structure interaction which are capable of theoretical analysis using essentially one-dimensional governing equations.

[1] Roberts, A. J. (1994). *A one-dimensional introduction to continuum mechanics.* World Scientific.

Simplification of more or less realistic problems leads to differential equations which can be readily solved: this is the essence of modelling with which engineers need to engage (and to realise that they are engaging) all the time. A few of these topics require some modest mathematical ability – a bit of integration, solution of ordinary and partial differential equations – but nothing beyond the eventual expectations of an undergraduate engineering degree programme. Sections that might, as a consequence, be omitted on a first reading, or until the classes in mathematics have caught up, are indicated by the symbol ♣.

Exercises are given for private study at the end of all chapters and some suggestions for laboratory demonstrations that could accompany such an introductory course are sprinkled through the book.

I am grateful to colleagues at Bristol and elsewhere – especially Danuta Lesniewska, Erdin Ibraim and Dick Clements – who have provided advice and comments on drafts of this book to which I have tried to respond. Erdin in particular has helped enormously by using material and examples from a draft of this book in his own teaching and has made many useful suggestions for clarification. However, the blame for any remaining errors must remain with me.

I am grateful to Christopher Bambridge, Ross Boulanger, Sarah Dagostino, David Eastaff, David Nash and Alan Powderham for their advice and help in locating and giving permission to reproduce suitable pictures.

I thank Bristol University for awarding me a University Research Fellowship for the academic year 2007–8 which gave me some breathing space after a particularly heavy four years of administrative duty.

I am particularly grateful to Peter Gordon for his editorial guidance and wisdom and his intervention at times of stress.

I acknowledge with gratitude Helen's indulgence and support while I have been preparing and revising this book.

David Muir Wood
Abbots Leigh
June 2009

SOIL MECHANICS

A one-dimensional introduction

1 Introduction

1.1 Introduction

In this chapter, we set the scene for the rest of the book. It may be helpful to remind readers of the relevance and importance of soil mechanics for all civil engineering construction: everything we construct sits on the ground in some way or other at some stage in its life. Even aircraft land on runways, and cars drive along roads; in each case there is some stiff layer (pavement) between the wheels and the prepared ground underneath. This stiff layer will help to spread the vehicular load but, in the end, this load must still be supported by the ground. Some examples of typical geotechnical design problems are presented in the next sections.

The term *soil mechanics* refers to the mechanical properties of soils; the term *geotechnical engineering* refers to the application of those mechanical properties to the design and construction of those parts of civil engineering systems which are concerned with the active or passive use of soils. Soils are the materials that we find in the ground: the term *ground engineering* is somewhat equivalent to *geotechnical engineering*. We will talk a little about the nature of soils in Chapter 3.

The term *soil* means different things to different people.[1] To an agricultural engineer, the soil is the upper layer of the ground which the farmer ploughs and harrows and in which crops are sown. To the civil engineer, this is the *topsoil*: it is recognised as valuable for agricultural purposes but usually too open in structure to be well suited for load bearing. Typically, the topsoil will be stripped from a construction site and stockpiled before serious construction activity begins. Underneath the topsoil is the soil with whose engineering properties we are concerned here. As we go further down into the ground, we will eventually encounter materials that no-one would have any hesitation in describing as *rock*. However, on the way there are many materials which can be described as hard soils or soft rocks: there is no precise

---

[1] Chambers' Dictionary defines *soil* as "the ground: the mould in which plants grow: the earth which nourishes plants" (from Latin *solum*, ground); but curiously *soil mechanics*, defined as "a branch of civil engineering concerned with the ability of different soils to withstand the use to which they are being put", is given as a sub-heading linked with Old French *souil*, wallowing place.

distinction and, although we are concerned primarily with soils in this book, much of what is presented will apply equally to weak rocks.

1.2 Soil mechanics

Soil mechanics is concerned with the mechanics of soils: a truism! It is an unfortunate feature of most civil engineering degree programmes that an atomistic approach to teaching is adopted: the teaching is broken down typically into structures, soils and water (with ancillary courses on mathematics, computing, and so on) and these subjects are developed in progressively greater detail through the three or four years of the degree with little opportunity being given to the embryonic engineers to integrate the separate elements and to recognise the importance of seeing the whole system as well as the parts. The whole may well combine elements that do not come from traditional civil engineering disciplines (mechanical or electrical, for example) and it is quite usual for systems to display emergent behaviours which had not been anticipated from the study of the constituent parts.

This book declares itself to be concerned with soil mechanics and therefore appears to perpetuate this segregation. However, opportunities will be taken where possible to demonstrate the behaviour of simple interactive systems: the interaction of soil with structural elements is a prime example where separate treatment of the different components is doomed to lead to erroneous expectations. But this book also attempts to integrate by addressing from a geotechnical point of view concepts which engineering students will be meeting as part of parallel courses on structures/materials or fluids/hydrostatics. Concepts of stress, stiffness, strength, fluid pressure, fluid flow and diffusion will, no doubt, be encountered in different contexts. The different treatments should not only indicate that there is a common thread of mechanics linking all aspects of engineering, but also encourage versatility and, further, provide reinforcement of learning.

1.3 Range of problems/applications

It is often said that the challenge of geotechnical engineering is that the soils have to be taken as they are found. Whereas structural materials such as steel and concrete can be designed to have particular desirable properties, the ground is as it is as a result of millennia of geological and geomorphological actions. The design of the geotechnical system will have to accommodate whatever properties the ground possesses. There may, on occasion, be some possibilities of modifying the *in-situ* properties of the ground to some extent. And where filling is required to build an embankment or a dam or to build up the ground behind a retaining wall, then it may be possible to consider this as a "designer soil".

From the point of view of design or analysis, the usual starting point with a geotechnical problem is to ensure that failure will not occur. Such stability or *ultimate limit state* calculations are often reasonably straightforward – certainly by comparison with the possibly more significant calculations of displacements under

1.3 Range of problems/applications

Figure 1.1. Military fortifications (Quebec City, Canada).

working conditions, the so-called *serviceability limit state* calculations. That there might be these two contrasting perspectives tells us that we will want to know about both strength and stiffness of soils – as of other materials.

Figures 1.1–1.13 show examples of geotechnical problems.

The design of military fortifications (Fig. 1.1) provided an important need to understand the way in which banks of earth could be supported by masonry walls. Coulomb, whose name is perhaps better known in the context of electrostatics, is known in soil mechanics as a pioneer of the practical analysis of earth pressure as a result of his experience in the design of fortifications for the French in Martinique in the eighteenth century. His challenge was to determine in a systematic way how large the masonry walls would need to be in order to guarantee the stability of the

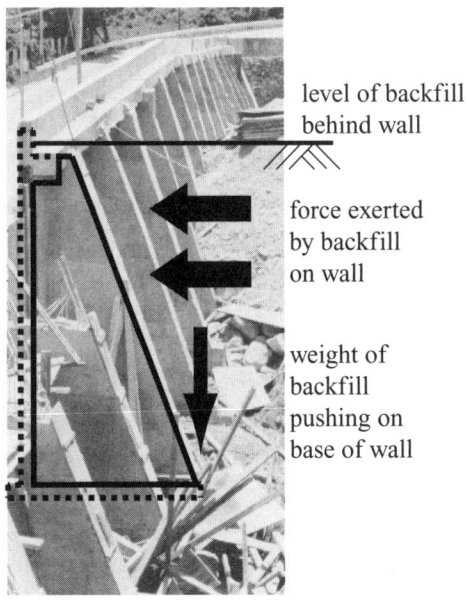

Figure 1.2. Retaining wall (New Territories, Hong Kong).

Figure 1.3. Deep excavation beside existing buildings (Century Hotel, San Francisco) (©Ross Boulanger).

ramparts. Nowadays retaining structures are often made of reinforced concrete and the shape of the wall can be chosen to make the supported soil – the *backfill* – help to stabilise the wall. The concrete cantilever retaining wall (Fig. 1.2) is being pushed over by the backfill, but the weight of the backfill is also pushing down on the base of wall and helping to stop the wall from overturning.

In congested urban areas, the value of land is so high that developers need to maximise the amount of space that they can generate and consequently build downwards as well as upwards. A deep excavation made beside existing buildings (Fig. 1.3) presents a more serious challenge of supporting the remaining ground in such a way that movements towards the excavated hole are kept small to avoid damage to the neighbouring structures. Tunnelling provides another means of exploiting underground space in cities for various sections of infrastructure – transport, drainage, services, flood relief (Fig. 1.4). Where tunnels are formed beneath existing buildings or near existing underground structures the deformations caused by ground loss during tunnelling must be kept small. The tunnel needs to be excavated and supported at the same time.

The retaining structures of Figs 1.1, 1.2, 1.3 are supporting the ground. Bridge structures rely on the ground for support. An arch bridge (Fig. 1.5) has to push laterally on the ground to generate the support for the dead and live load that it

1.3 Range of problems/applications

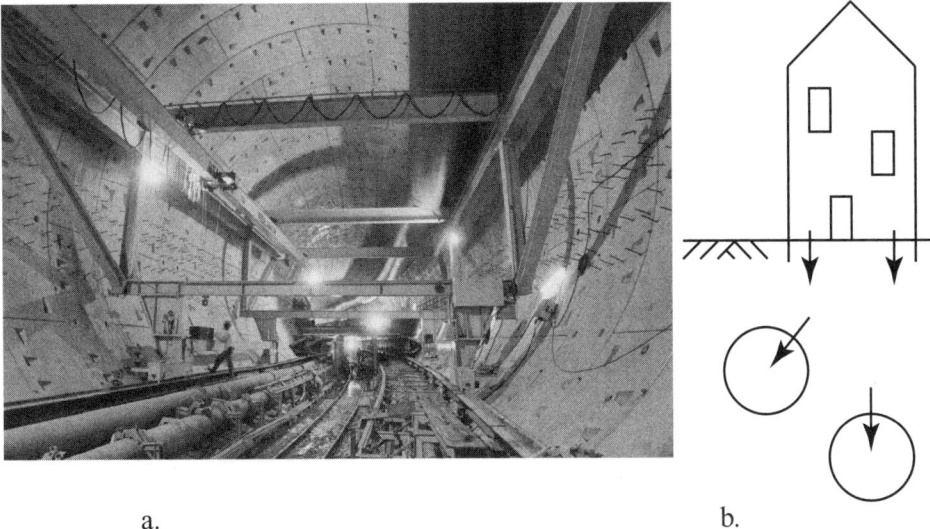

a.   b.

Figure 1.4. (a) Combined flood relief and road tunnel under construction (Kuala Lumpur, Malaysia) (©Mott MacDonald); (b) tunnelling beneath existing foundations.

carries: if the lateral push cannot be sustained then the bridge may deform in a way which is not comforting to the public. The bridge will expand and contract with seasonal variation of temperature and, as it expands, the supporting abutments will be pushed against the soil. What will be the forces on these abutments? We will need to design the structural elements to take account of the interaction between the abutment and the ground. Traffic on the roads will provide many repeated cycles

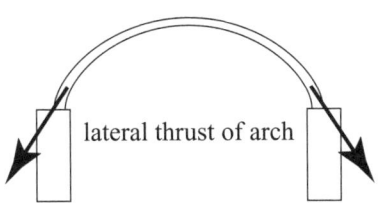

Figure 1.5. Arch bridge (Constantine, Algeria).

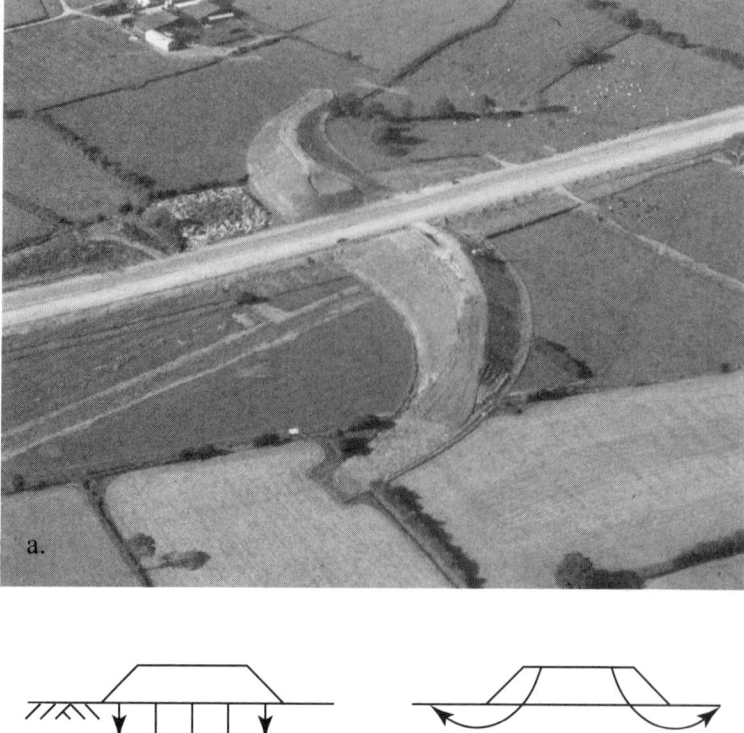
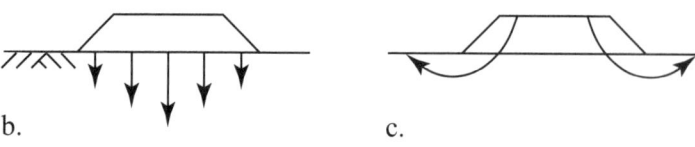

Figure 1.6. (a) Embankments on soft ground (M4 motorway and overbridge approach embankments, Bristol, UK) (©Sarah Dagostino); (b) settlement of embankment; (c) failure of side slopes of embankment.

of loading on the ground. The foundations of roads are usually engineered to ensure that the deformations of the road surface are tolerable.

On coastal or estuarine sites, it is often necessary to construct roads on soft ground, perhaps on man-made embankments (Fig. 1.6) to ensure that the road is above any likely level of inundation. Soft ground by definition is expected to show significant movements as it responds to the applied embankment load (Fig. 1.6b). It may not be possible to prevent such movements but they will need to be understood and managed in order to prevent failure of the embankment (Fig. 1.6c). Soft ground may not provide adequate support for shallow foundations (Fig. 1.7). The deformations may eventually become too much for the continued use of the building. A concrete platform for offshore oil production (Fig. 1.8) will eventually sit on the seabed and may store oil or water as ballast in integral concrete tanks; the deck provides operational and living quarters. The structure is subjected to all the power of the sea, which will try to overturn it or shift it sideways. Pile foundations (Fig. 1.9) are a means by which the loads from a highway structure can be taken deep into the ground away from other structures. By transferring the loads to firm soils or rocks

1.3 Range of problems/applications

Figure 1.7. Row of cottages founded on soft ground: (a) before (1973) and (b) after collapse/demolition of end cottage (1977). (Stanford Dingley, UK).

at depth the potential problems of settlement or failure of softer, near-surface soils (Figs 1.6, 1.7) are eliminated.

Vertical and horizontal shaking of the ground by an earthquake provides another form of loading which may need to be considered. Some sandy soils lose their strength as they are shaken and turn into quicksand as they liquefy – a seaside effect with which most children are familiar. The apartment blocks founded on such a soil at Niigata (Fig. 1.10) failed by rotation but remained largely structurally intact as they rotated.

In most of these examples, there will be concerns about the movements that may occur but in addition, in Figs 1.3, 1.4, 1.5 and 1.9, there will be concern about the consequences for the structural elements (retaining structure in Fig. 1.3; tunnel

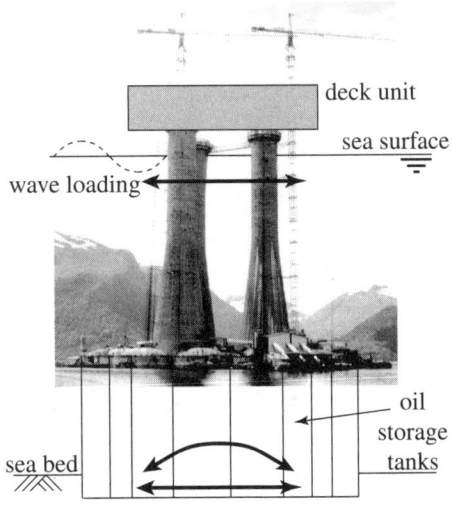

Figure 1.8. Concrete oil production platform under construction (Åndalsnes, Norway).

piles to transfer loads from highway
columns to strong ground at depth

Figure 1.9. Highway structure founded on piles (Kowloon, Hong Kong).

or culvert linings in Fig. 1.4; bridge abutments in Fig. 1.5; pile foundations in Fig. 1.9; existing buildings in Figs 1.3, 1.4) of these movements. These are certainly examples where potential soil-structure interaction issues will need to be considered.

It will become clear that water plays an important part in the mechanics of soils and geotechnical engineering. It may be necessary to create dry excavations within or near a lake, river or sea protected by a cofferdam of driven sheet piles as sketched in Fig. 1.11a. Sheet piles are driven into the ground to form an enclosed space protecting the excavation from inflow of water and provide an environment

Figure 1.10. Failure of apartment blocks at Niigata, Japan as result of liquefaction induced by earthquake in 1964.

1.3 Range of problems/applications

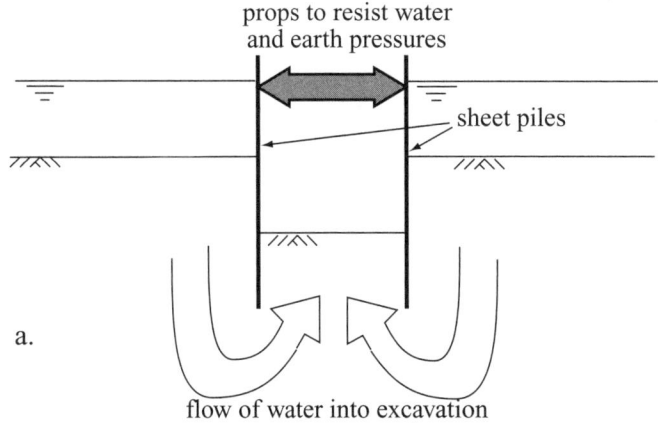

Figure 1.11. (a) Sheet pile cofferdam; (b) struts propping sheet pile walls to protect excavation.

within which the construction of the foundations for the pier of a bridge, or the laying of a pipeline, can proceed. The water and the soil around the piles will be pushing in: the forces must be resisted, for example, by props or struts near the top of the piles (Fig. 1.11b) – this is just another form of retaining structure. The control of the flow of water from the surroundings into the base of the excavation is a classic geotechnical design situation: pumps will be needed to keep it dry and we will need to be able to calculate the pumping capacity that is required.

An earth- or rock-fill dam may be used to form a reservoir (Figs 1.12 and 1.13). Such a dam is really just an enlarged and more sophisticated version of the hydraulic structures that we or our children build at the seaside. There is a core of some impermeable material providing a barrier, for example, clay (Fig. 1.12), or asphaltic concrete (Fig. 1.13). The core is intended to restrict the flow, but it may be impossible to eliminate leakage altogether. More permeable materials – graded crushed rock and rock-fill – are placed upstream and downstream as transition zones to

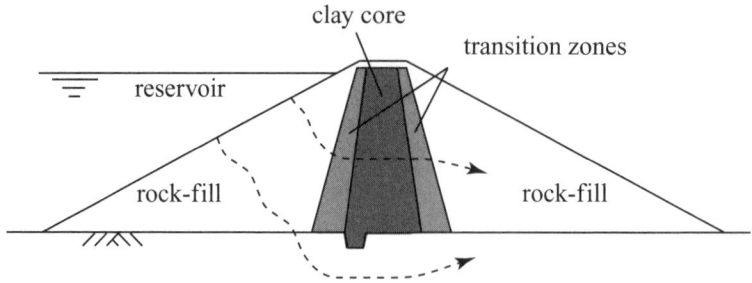

Figure 1.12. Rock-fill dam.

provide protection to the core, and discourage erosion. Such a geotechnical system is a quintessential application of "designer soils" where each of the zones of the dam will consist of soils chosen for their specific properties.

1.4 Scope of this book

A one-dimensional approach to these real-world problems involving soils may seem somewhat far-fetched but it will be seen that, for many of them, although the overall geometry is certainly three-dimensional, much of the detail of the mechanical

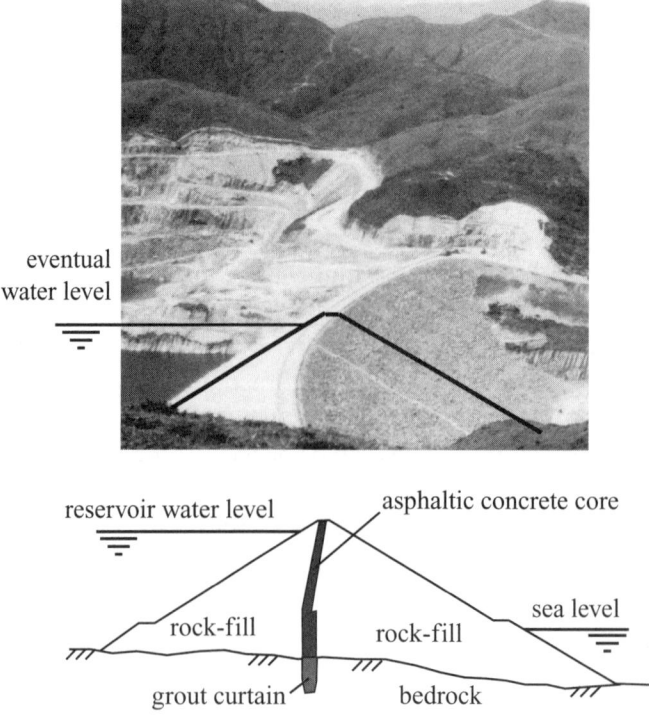

Figure 1.13. Dam with core of asphaltic concrete (High Island, Hong Kong).

response of the soil as a material does not require anything more than the one-dimensional model in order that it should be understood. There may be geometrical effects associated with collapse mechanisms which exploit the second and third dimensions of the problem but the soil response can be adequately incorporated using a simple model.

However, this book is not meant to be an apologia for adopting such a one-dimensional approach. Rather, it uses this approach to gather information and look for more or less consistent patterns which can then aid our subsequent development of more complete models – adding the extra dimensions as we go.

1.5 Mind maps

Students (and others) have found that the use of *mind maps*[2] can be very helpful for the recording and recalling of bodies of knowledge, and for fixing the interactions between different segments of that knowledge. Rudimentary mind maps have been produced at the end of each chapter to summarise what are regarded as the key points that have been introduced in that chapter. These mind maps are intended merely to be illustrative – students are advised to construct their own mind maps in order that they may be confident that they have imposed their own mental processing on the content.

[2] Buzan, T. (1974). *Use your head.* BBC Books.

2 Stress in soils

2.1 Introduction

In this chapter, we introduce the concept of stress and demonstrate how we can calculate stresses in the ground, recalling that we are only concerned with configurations that can be described as one-dimensional. Initial sections rehearse some of the ideas of mechanics – Newton's First Law, the distinction between mass and weight, and the idea of gravity. The single dimension of our problems allows us to impose some notions of symmetry which are helpful in simplifying our calculations of stresses in soils.

We need then to introduce the possible presence of water in the ground. Some background discussion of basic hydrostatics is required in order that we may describe the pressures that exist in the water. We end with a powerful hypothesis about the way in which stresses are shared between the water and the soil.

2.2 Equilibrium

Newton's First Law of Motion says that an object will remain in its state of rest or of uniform motion in a straight line unless acted upon by an out-of-balance force.[1] We need not concern ourselves here with the possibility of motion – our soils are expected to be rather stationary or at least to move only very slowly as a result of some construction process. The condition of rest or stasis therefore requires that the forces acting on an object should be in balance.

Evidently forces can be imposed on a completely general object from all directions (Fig. 2.1a) and Newton's Law requires that this general set of forces should be in complete balance in order that the object should remain stationary. However, we are restricting ourselves to a single dimension (Fig. 2.1b) and Newton's Law says that the sum of all the n forces F_{zi} in this single direction – let us call it the z direction – should be zero:

$$\sum_{i=1}^{n} F_{zi} = 0 \qquad (2.1)$$

---

[1] Lex I: *Corpus omne perseverare in statu suo quiescendi vel movendi uniformiter in directum, nisi quatenus a viribus impressis cogitur statum illum mutare.*

2.3 Gravity

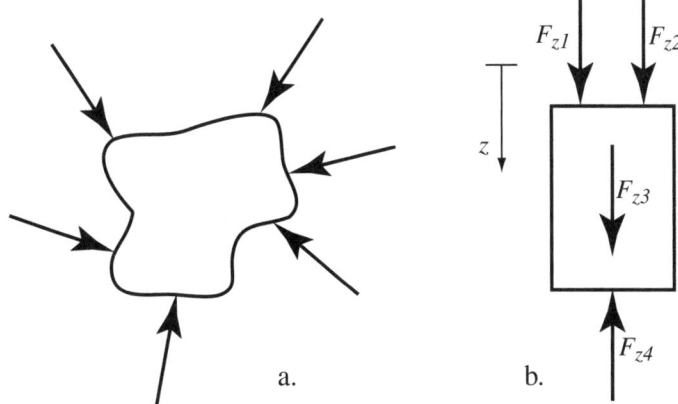

Figure 2.1. (a) Object subjected to general system of forces; (b) object subjected to one-dimensional system of forces in z direction.

Applying this to the schematic example of Fig. 2.1b, this becomes:

$$F_{z1} + F_{z2} + F_{z3} - F_{z4} = 0 \qquad (2.2)$$

We note that the forces shown in Fig. 2.1b have one of two directions: up or down. The direction in which forces act is important and we must be careful to assign the appropriate positive or negative sign to the individual forces in (2.2). Thus the forces F_{z1}, F_{z2}, F_{z3} are all acting in the positive, downward, z direction and appear with positive sign in (2.2), whereas force F_{z4} acts in the negative, upward, $-z$ direction and therefore appears with a negative sign in (2.2).

This description of static equilibrium seems intuitively obvious. Imagine two (or three) children fighting for possession of a favourite toy. They each pull as hard as they can but they are equally matched, the forces balance, and the toy does not move. (Eventually, the toy may break so that each of the children obtains a portion but that is to do with *strength*, which is the subject of Chapter 8.)

2.3 Gravity

What are the principal sources of forces on an object, which in the present context will be a chunk of soil in the ground? There will generally be forces which arise because of the actions of man – buildings or other civil engineering works – but a primary source of forces in the ground arises from the self-weight of the material and the gravitational pull that the chunk of soil feels towards the centre of the Earth. This is an example of a *body force* which acts throughout the material: F_{z3} in Fig. 2.1b is intended to represent such a force, which is always present no matter what else may be happening.

The Law of Gravitational attraction is also something for which Newton was responsible. It proposes an inverse square law to describe the force mutually

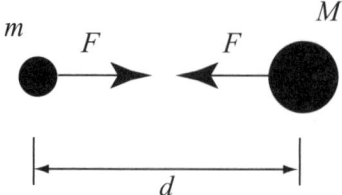

Figure 2.2. Newton's Law of Gravitational attraction: the force F mutually experienced by two masses m and M whose centres are at a distance d apart.

experienced by two point masses[2] m and M whose centres are at a distance d apart (Fig. 2.2):

$$F = G\frac{mM}{d^2} \quad (2.3)$$

where G is the gravitational constant with the value 6.67×10^{-11} N-m^2/kg^2. If one or both of the masses has a dimension that cannot be ignored, then we can break that mass into a large number of infinitesimally small masses and apply the Law of Gravitational attraction to each of the constituent parts. We will find that the result is equivalent to taking d as the distance between the centres of mass of the two objects.

We are concerned with effects on the surface of the Earth. The radius of the Earth is (on average) about 6.38×10^6 m and cannot be ignored, although the dimensions of any geotechnical engineering system are likely to be very small in comparison. We can assign the second mass M in (2.3) to the mass of the Earth (5.98×10^{24} kg) so that the gravitational force exerted on a mass m by the Earth can be written simply as:

$$F = mg \quad (2.4)$$

where $g = 9.81$ m/s^2 is the acceleration due to gravity and can be deduced from the values of the mass and radius of the Earth and the gravitational constant G.

An object dropped in free air falls to the ground at an increasing velocity because it is subject to the gravitational force of attraction which, being an out-of-balance force, causes the object to enter a state of *non-uniform* motion and hence to accelerate, with the acceleration due to gravity $g = 9.81$ m/s^2 (Fig. 2.3).[3] An object sitting on the ground is still subject to the gravitational attraction of the Earth but this is resisted by the ground: there is an equal and opposite force exerted by the ground on the object so that the resulting acceleration is zero. Thus, in Fig. 2.4, the

---

[2] A *point mass* is a nice concept: all the mass is packed into a volume of zero dimension. All it is really saying is that the finite dimensions of the masses are small by comparison with their separation. This is a pragmatic modelling assumption.

[3] According to Newton's Law of Gravitational attraction, the Earth is also experiencing an attractive force F towards the object but the resulting acceleration is negligible because of the disproportion of the masses of the Earth and the object (unless the object is the Moon or some other massive celestial body).

2.3 Gravity

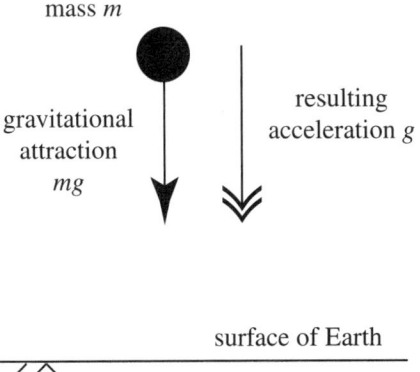

Figure 2.3. An object of mass m subject to gravitational attraction mg accelerates towards the Earth with acceleration $g = 9.81$ m/s^2.

object of mass m and the force F exerted by the ground on the object are related by Newton's First Law (2.1):

$$F - mg = 0 \tag{2.5}$$

or

$$F = mg \tag{2.6}$$

Equation (2.6) allows us to emphasise the difference between mass and weight. The *mass* of an object is quite simply the amount of matter in the object. However, the *weight* of an object is the force exerted by gravity on the object. Mass is independent of gravitational field – the mass of an object will be the same on the surface of the Moon or on the surface of a planet as it is on the surface of the Earth. However, the gravitational pull on the surface of each celestial body towards the centre of that body will differ so that the *weight* of the object will change (Fig. 2.5). Mass has units of kilograms (kg); weight is a force and has units of newtons (N). One newton is

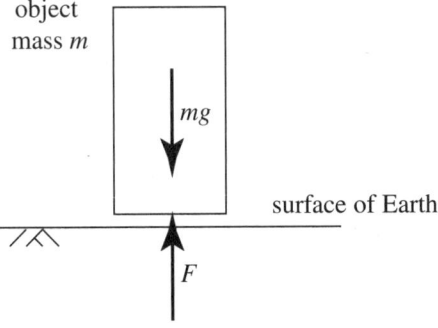

Figure 2.4. An object of mass m resting on the Earth's surface is subject to gravitational attraction mg but is unable to accelerate, so must be resisted by a force $F = mg$ exerted by the ground on the object.

Figure 2.5. The Little Prince on the surface of his Asteroid B-612 probably experiences considerably less gravitational attraction than when he appears on the surface of the Earth (unless the asteroid has an extremely large mass). (Saint Exupéry, Antoine de (1943). *Le Petit Prince*. ©Editions Gallimard).

the force required to give a mass of 1 kg an acceleration of 1 m/s^2. On the Earth's surface, a mass of 1 kg will thus have a weight of 9.81 N $\approx$ 10 N. On the moon the acceleration due to gravity is about 1.62 m/s^2 and the mass of 1 kg will correspondingly weigh only 1.62 N.

2.4 Stress

The application of Newton's First Law to the equilibrium of an object was necessarily presented in terms of the *forces* acting on that object. It is a matter of experience that the spreading of a force over a larger area will help to reduce its effects or the damage that it causes. Readers of Arthur Ransome's east coast tale *Secret water*[4] will recall that a character nicknamed the Mastodon was able to move around on the soft mud-flats by attaching tennis racket-like contraptions to his feet – essentially, classic snowshoes (Fig. 2.6). The effect of these shoes was on the one hand to leave mysterious footprints on the mud and on the other to spread the weight of the body over a larger area. The intensity of the force over an area is called *stress*.

Some simple calculations can be made. The mass of a typical human might be around 75 kg. Estimating the area of a typical foot at, say, $0.2 \times 0.08 = 0.016$ m^2 then the stress imposed on the ground by the person standing on one foot would be $75 \times 9.81/0.016 = 46,000$ N/m^2. The unit of stress, newtons per square metre, is called a pascal. In soil mechanics the stresses that we will encounter are generally of the order of thousands (or even millions) of pascals and it is more convenient to work

---

[4] Ransome, A. (1939). *Secret water*. Jonathan Cape.

2.4 Stress

Figure 2.6. The "Mastodon" using "splatchers" to spread the load on soft mud. (Ransome, Arthur (1939) *Secret water* Jonathan Cape. Reprinted by permission of The Random House Group Ltd.)

in terms of kilopascals (kPa) (or megapascals (MPa)). The contact stress imposed by this human is thus about 46 kPa. If we spread the weight over "mud-shoes" of diameter about 0.25 m, with individual area 0.049 m^2, then the stress required to support the human standing on two feet is reduced to 7.5 kPa.

An elephant is a large creature (Fig. 2.7a) and we might expect that the stresses beneath the feet of an elephant would be significantly higher than those under the human feet. The mass of an elephant is typically around 4500 kg (60 times the mass of our human). If we guess that the diameter of a typical elephant foot is around 0.4 m (area 0.126 m^2), then, an elephant standing on all four feet generates a contact stress of around 88 kPa, only about twice the stress supporting our human standing on one foot. However, if our human is wearing shoes with narrow heels (Fig. 2.7b), with contact diameter of perhaps 0.01 m (area 78×10^{-6} m^2), then the contact stress, assuming that the person is standing only on the two heels, is some 4700 kPa or 4.7 MPa – where 1 MPa or one megapascal is one thousand kilopascals or one million pascals. So, in terms of damage to your floor (and setting aside

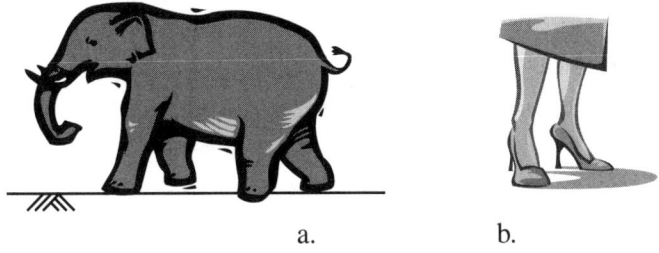

Figure 2.7. (a) Elephant and (b) small heels.

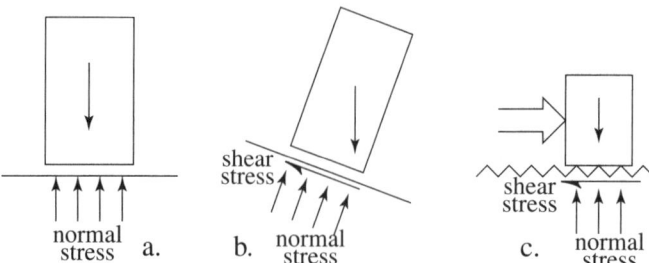

Figure 2.8. (a) Block supported by normal stress; (b) block on inclined surface supported by normal stress and shear stress; (c) block resisting sideways push mobilising shear stress and normal stress.

other possible practical disadvantages) you would do better to invite into your home an elephant than an elegant lady wearing stiletto heels. This is something of which owners of stately homes are well aware (though they probably avoid the elephants too).

The stresses that we have been calculating here are *normal* stresses, directed orthogonally to the interface (Fig. 2.8a), and in thinking about the mechanics of soils we usually assign a positive sign to stresses that are tending to *compress*, to push elements of the system together.[5] That the force required to support an object and prevent it from moving (Newton's Law again) should have the freedom to take any direction necessary to maintain equilibrium seems quite reasonable: thus with an object supported on an inclined surface (Fig. 2.8b) there will be forces and hence stresses both normal to the interface and parallel to the interface. The latter are called *shear* stresses. We will meet shear stresses in Chapter 8 when we consider the strength of soils within our one-dimensional framework. Figure 2.8c shows a schematic diagram of a block resisting a sideways push on a rough interface. Again, considerations of equilibrium suggest that there will be a need for shear stresses parallel to the interface as well as normal stresses supporting the weight of the block.

2.5 Exercises: Stress

1. What is the weight of an object of mass 400 kg?
2. A man of mass 80 kg has shoes (UK size 12, US size $12\frac{1}{2}$, European size 47) with a contact area with the ground of about 0.025 m² for each foot. What is the contact normal stress when he stands on both feet?
3. Study of dinosaur footprints suggests that *Tyrannosaurus rex* may have exerted a stress on the ground of about 120 kPa. From study of dinosaur fossils it is estimated that the total area of the feet of *Tyrannosaurus rex* was about 0.6 m². What was the mass of this dinosaur?

---

[5] In other branches of engineering mechanics, it is common to treat as positive tensile stresses which are tending to pull things apart. The sign convention adopted may or may not be explicitly specified so *caveat lector*, be alert to the potential for confusion.

2.6 Vertical stress profile

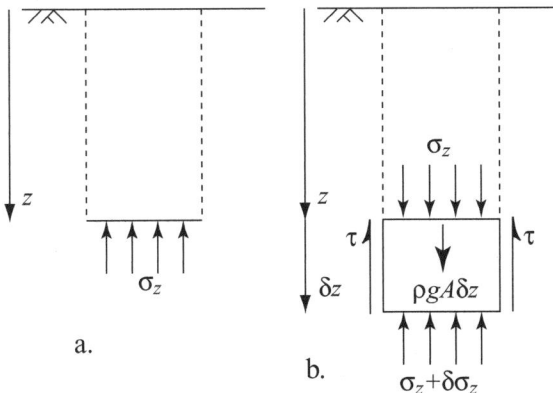

Figure 2.9. (a) Stress σ_z supporting column of soil at depth z; (b) element of column of soil in ground between depths z and $z + \delta z$.

4. A man of mass 90 kg wearing a space-suit with area of each foot equal to 0.06 m² lands on the Moon and starts walking over the lunar landscape. What is the contact stress when he stands on one leg to adjust the buckle on the other boot?
5. What diameter of circular "snow shoes" on each foot would be necessary in order to reduce the contact stress of the 80 kg man in Question 2 to 5 kPa when standing on both feet?
6. The Little Prince's Asteroid B-612 (Fig. 2.5) appears to have a diameter of about 4.5 m. Let us suppose that he has a mass of about 50 kg and that the asteroid is made of the same material as the Earth. What is the gravitational force of attraction of the Little Prince to his asteroid?

2.6 Vertical stress profile

We are aiming to calculate stresses in the ground and, true to our one-dimensional constraint, we are going to consider configurations which are described by a single spatial coordinate: depth z below the ground surface.[6] The ground profile is thus assumed to be independent of lateral position, at least at the scale at which we are operating. We will start by considering the stress needed to support a column of soil in the ground as shown in Fig. 2.9.

First let us assume that the stress in the ground at the base of a soil column at depth z is σ_z (Fig. 2.9a). We now take an element out of the soil column, with cross-sectional area A, between depths z and $z + \delta z$ (Fig. 2.9b). Equilibrium tells us that the stress σ_z holding up the soil column above depth z must also be bearing down on the top of the element. The stress, $\sigma_z + \delta \sigma_z$, at the base of the element can then be deduced from vertical equilibrium of the element, applying (2.1). The forces on

---
[6] Just as sign conventions for stress need a watchful eye, so do sign conventions for vertical position. In general, we will start at the ground surface and measure depth z positive *downwards*. However, there may be occasions (for example, in analysing flow of water, Chapter 5), where it is really more helpful to measure height above some reference datum so that the coordinate z will be positive upwards. We will draw attention to that choice when it is made.

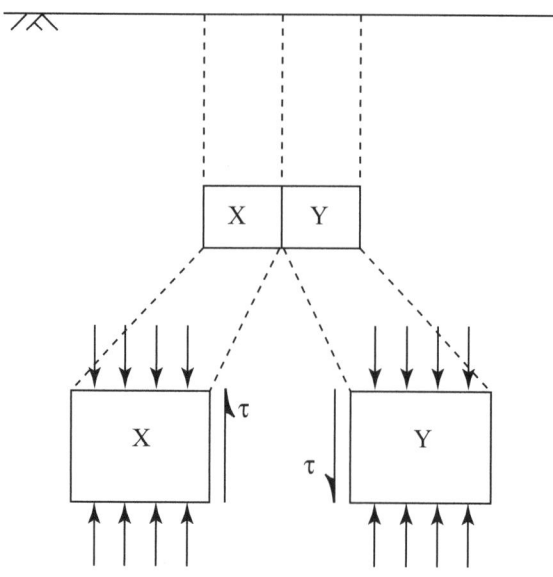

Figure 2.10. Symmetry argument for deducing that shear stress $\tau = 0$.

the element are the stress σ_z acting over area A at the top of the element, the self-weight of the element calculated from its volume $A\delta z$, density ρ and gravitational acceleration g, together with any shear stresses τ acting round the periphery of the element, which has an area $s\delta z$ where s is the perimeter. We find then:

$$-(\sigma_z + \delta\sigma_z)A + \sigma_z A + \rho g A \delta z - \tau s \delta z = 0 \qquad (2.7)$$

What can we deduce about the shear stress τ?

There are often arguments that can be used to simplify analyses which rely solely on appreciation of the symmetry constraints of the problem. We are considering the concept of a soil layer of large lateral extent. This implies that conditions at the same depth z everywhere in the soil layer are the same. Let us suppose that there are indeed non-zero shear stresses on element X in one column (Fig. 2.10). Equilibrium, in the sense of balancing of forces across interfaces, requires that there must be an equal and opposite shear stress on the adjacent element Y. But the soil mass is of large lateral extent and no one soil column is any different from any other soil column: it is not possible for there to be shear stresses in opposite directions on identical columns. The only permissible solution is that the shear stresses on vertical boundaries are in fact zero. Another way of thinking through this question is to imagine viewing the pair of elements X and Y from the back (Fig. 2.11): the signs of the shear stresses are reversed but, since the choice of elements was entirely arbitrary, there can be no difference between the elements in Fig. 2.10 and Fig. 2.11: we could interchange the labelling of the elements X and Y. The only way to make the elements in these two figures match is to insist that the interface shear stress τ between the elements is zero. (An exactly similar argument could be used to demonstrate that the stress on the horizontal base of any section through our soil

2.6 Vertical stress profile

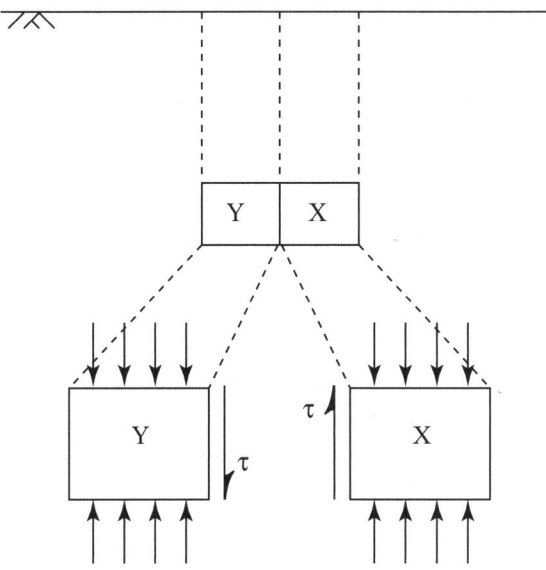

Figure 2.11. Symmetry argument for deducing that shear stress $\tau = 0$: elements of soil from Fig. 2.10 seen from the back.

column must be solely a normal (vertical) stress with no shear (horizontal) component.)

Setting $\tau = 0$ and rearranging (2.7) we find:

$$\delta\sigma_z = \rho g \delta z \qquad (2.8)$$

and we can integrate this to give:

$$\sigma_z = \int_0^z \rho g \, dz \qquad (2.9)$$

For a uniform soil with density ρ constant throughout:

$$\sigma_z = \rho g z \qquad (2.10)$$

and for a layered soil with n layers of thickness Δz_i and density ρ_i:

$$\sigma_z = \sum_{i=1}^{n} \rho_i g \Delta z_i \qquad (2.11)$$

2.6.1 Worked examples

1. The soil at a certain site is uniform down to great depth with constant density $\rho = 2$ Mg/m^3. Calculate the vertical stress at depths of 2, 5 and 8m.

   For such a uniform soil the vertical stress at any depth is given by (2.10). With density $\rho = 2$ Mg/m^3 the stress at depth z is $\rho g z = 19.62z$ kPa (with z measured in metres). It is important to keep a close watch on the units in any calculation. One newton is the force required to give a mass of 1 kg an acceleration of

1 m/s². If you work consistently in kilograms and metres, calculations of stresses will emerge in pascals where one pascal (1 Pa) is equal to one newton per square metre (1 N/m²). However, it may often be convenient to deal with recognised multiples of these basic units such that the numbers themselves can be written more conveniently (though of course the magnitudes of the quantities do not change). Thus densities might be written in megagrams (Mg) rather than kilograms (kg) per cubic metre: density of water is 1 Mg/m³ – or 1000 kg/m³; density of soil is typically around 1.5–2 Mg/m³ – or 1500–2000 kg/m³. Typical stresses in the ground are in the thousands of pascals, as we see here: it makes sense to use kilopascals as the unit. Thus in using the given information about density in Mg/m³ and producing a result for vertical stress in kPa we have incorporated two cancelling factors of 1000 in this expression. We have defined our density ρ in terms of megagrams (1 megagram = 1000 kilograms) and have produced a result in kilopascals (1 kilopascal = 1000 pascals). Beware!

At depth 2 m, $\sigma_z = 39.24$ kPa, at depth 5 m, $\sigma_z = 98.1$ kPa, and at depth 8 m $\sigma_z = 156.96$ kPa.

It is often convenient to approximate the acceleration due to gravity as 10 m/s² which makes mental arithmetic calculation slightly simpler and the numbers rounder. The three stresses in this example would then become 40, 100 and 160 kPa at depths 2, 5 and 8 m, respectively.

2. Figure 2.12 shows a layered soil profile. Calculate the vertical stress at the base of each layer.

At the ground surface, the vertical stress is $\sigma_z = 0$.

At a depth of 2 m, at the base of the layer of silty sand with density $\rho = 2.1$ Mg/m², the vertical stress is $\sigma_z = 2 \times 2.1 \times 9.81 = 41.2$ kPa.

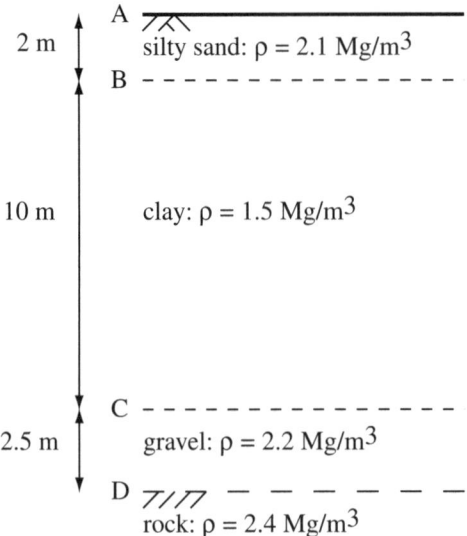

Figure 2.12. Profile of layered ground.

2.7 Water in the ground: Introduction to hydrostatics

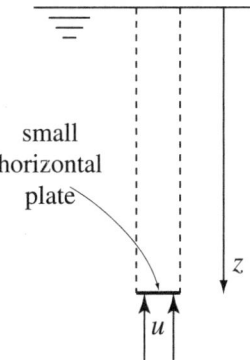

Figure 2.13. Pressure in pool of water acting on small horizontal pressure plate.

At a depth of 12 m, at the base of the clay layer with density $\rho = 1.5$ Mg/m^3, the vertical stress is the sum of the effects of the layer of silty sand and the layer of clay: $\sigma_z = 2 \times 2.1 \times 9.81 + 10 \times 1.5 \times 9.81 = 41.2 + 147.1 = 188.3$ kPa.

At a depth of 14.5 m, at the base of the gravel layer with density $\rho = 2.2$ Mg/m^3, the vertical stress is the sum of the effects of the layer of silty sand and the layer of clay and the layer of gravel: $\sigma_z = 2 \times 2.1 \times 9.81 + 10 \times 1.5 \times 9.81 + 2.5 \times 2.2 \times 9.81 = 188.3 + 54.0 = 242.3$ kPa.

2.7 Water in the ground: Introduction to hydrostatics

If you dig a hole on the beach then it is likely that you will eventually encounter water because of the proximity of the sea. In fact in temperate climates (such as the United Kingdom) there is quite a good chance that any hole will encounter water if it is dug deep enough. The height above sea level at which this *water table* is encountered will depend on local topography and ground conditions. We need to understand how the presence of water will influence our thoughts about the stresses existing in the ground. We therefore introduce a brief discussion about hydrostatics.

Hydrostatics is that branch of mechanics which is concerned with fluids which are stationary. We will encounter (slowly) moving fluids in Chapter 5. For the most part, the fluid with which we will be concerned will be water – though there may be situations where there are hydrocarbons or other liquids present. We start by thinking of a pool of water – a swimming pool or a lake – with no soil present. The first idea to be grasped is that of fluid pressure. Following an argument similar to that used to calculate the vertical stress in soils in Section 2.6 we can deduce that the stress acting on a little horizontal plate at a depth z below the surface of the pool (Fig. 2.13) will be:

$$u = \rho_w g z \tag{2.12}$$

where ρ_w is the density of the water.

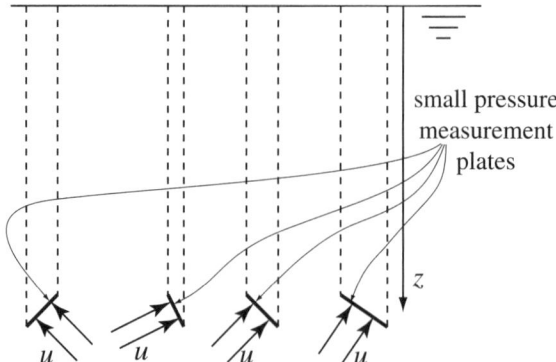

Figure 2.14. Pressure in pool of water acting on small pressure plates at different orientations.

The density of fresh water is $\rho_w = 1$ Mg/m^3 = 1000 kg/m^3. However, the dissolved salts in sea water increase the density. Typical ocean water has a salt content of about 3.5% and a density of around 1.027 Mg/m^3. The Dead Sea has a salt content of around 35% and density 1.24 Mg/m^3, thus making it easier to float. By contrast, a somewhat enclosed sea like the Baltic, with much regular injection of fresh water, has a salt content typically around 1% and density only slightly greater than fresh water.

It is a property of fluids at rest that the stress exerted is independent of the orientation of the little measuring plate on which it is acting but is dependent only on the depth below the free surface (Fig. 2.14). Stationary water cannot transmit shear stresses. We can then imagine that these little measuring plates are in fact just measuring planes drawn in the fluid but having no material presence: the resulting stress on each plane is, as before, the same. Such a "hydrostatic" stress in a fluid is more usually described as a fluid *pressure* and, since we are not usually concerned to try to pull fluids apart, pressure is helpfully deemed to be positive in compression.

Another way of visualising this intuitive property is to imagine standpipes with various geometries inserted in the pool at the same depth as shown in Fig. 2.15. The water in each standpipe will rise to the level of the water surface. The pressure at the mouth of each standpipe is found directly from the height z of the free surface above the opening.

We know that objects float in water. Archimedes' principle says that an object wholly or partially immersed in a fluid (water will be a particular fluid) is buoyed up by a force equal to the weight of the displaced fluid. We can understand this result by applying our recently acquired knowledge of pressure in fluids. Let us consider a cuboidal object wholly submerged in the pool of water with the faces of the cuboid aligned with vertical and horizontal axes, as shown in Fig. 2.16. The lengths of the sides of the object are Δx, Δy and Δz and the top surface of the object is at a depth z below the water surface. There will be forces F_{x1}, F_{x2}, F_{y1}, F_{y2}, F_{z1}, F_{z2} acting on the faces of the object, as shown. We can calculate the magnitude of each of these forces by integrating the water pressure over each face. Because of the chosen

2.7 Water in the ground: Introduction to hydrostatics

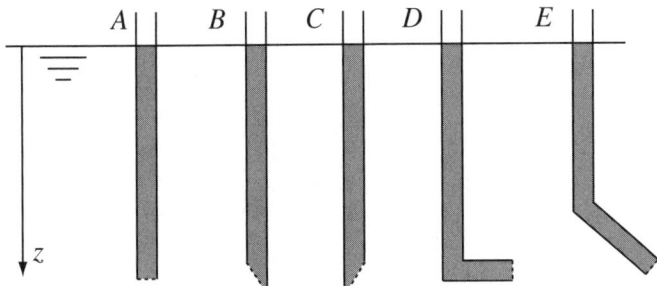

Figure 2.15. Standpipes A, B, C, D, E in pool of water with opening at depth z: the water rises to exactly the same level in each of the standpipes.

orientation of the object, the forces on opposite vertical faces must balance exactly: $F_{x1} = F_{x2}$, $F_{y1} = F_{y2}$. The water pressure acts perpendicularly to the face and for every little element on one face acted on by a certain magnitude of water pressure there is an identical element on the opposite face acted on by the same magnitude of water pressure. The water pressure varies linearly with depth and hence the average pressure on each of the faces of the element shown in Fig. 2.16 can be calculated at the midpoint of that face. Thus $F_{x1} = F_{x2} = \rho_w g(z + \Delta z/2)\Delta y \Delta z$; $F_{y1} = F_{y2} = \rho_w g(z + \Delta z/2)\Delta x \Delta z$. (These forces act on the faces of the cuboidal element at a level $z + 2\Delta z/3$ below the surface of the pool.)

In the vertical, z, direction we can calculate the force on the top of the object from the pressure $-\rho_w g z$, from (2.12) – and the area over which it acts ($\Delta x \Delta y$):

$$F_{z1} = \rho_w g z \Delta x \Delta y \qquad (2.13)$$

Similarly, the force on the bottom of the object can be calculated as:

$$F_{z2} = \rho_w g(z + \Delta z)\Delta x \Delta y \qquad (2.14)$$

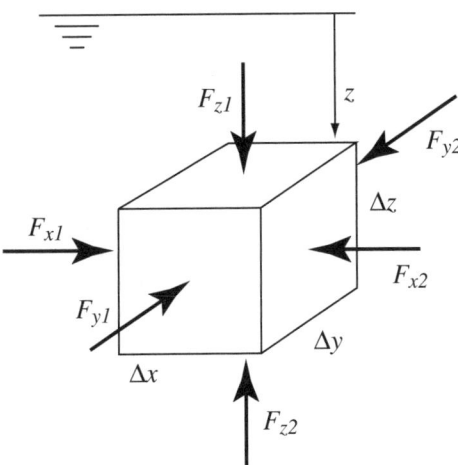

Figure 2.16. Cuboidal object submerged in a pool of water: water pressure produces forces $F_{x1}, \ldots$ on each face of the object.

Figure 2.17. Hot air balloon taking advantage of Archimedes uplift.

The out-of-balance force tending to push the object upwards is:

$$F_{z2} - F_{z1} = \rho_w g(z + \Delta z)\Delta x \Delta y - \rho_w g z \Delta x \Delta y = \rho_w g \Delta z \Delta x \Delta y \qquad (2.15)$$

and this is precisely the weight of the water displaced by the object. This is an out-of-balance force from the water pressures: the object will tend to move up or move down or stay where it is depending on its own weight in relation to the weight of the water that it displaces. A submerged object with a weight less than the weight of the displaced water will have to be tethered in some way to maintain its position; if it has a greater weight then it will have to be propped if it is not to sink. A hot air balloon has neutral buoyancy and floats in its surrounding fluid – air – with its weight exactly matching the Archimedean uplift (Fig. 2.17). The density of the Earth's atmosphere decreases with height above the Earth's surface so, to enable the balloon to climb, the air in the balloon is heated to decrease its density and hence its weight: the balloon rises until a vertical equilibrium is again attained.

Our calculation of the buoyancy force on the cuboidal object in Fig. 2.16 can be applied to an element of any size – we can make Δx, Δy and Δz as small as we like, even infinitesimally small. Any object of any shape can be thought of as made up of a large number of very small elements and, applying principles of calculus, we can integrate the out-of-balance force to discover that it is always equal to the weight of the water that is displaced.

2.7.1 Worked example: Archimedes uplift on spherical object

1. A solid sphere of aluminium alloy with diameter 50 mm is suspended from a wire in a bowl of fresh water (Fig. 2.18). What is the force F_1 in the wire?

2.7 Water in the ground: Introduction to hydrostatics

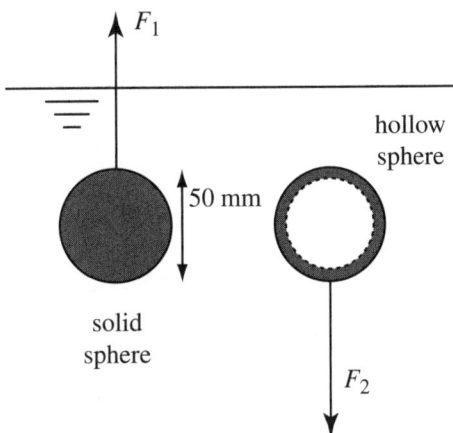

Figure 2.18. Solid and hollow spheres suspended in fresh water.

The aluminium alloy has a specific gravity of 2.7 and hence a density of $\rho_a = 2.7$ Mg/m^3.

The volume of a sphere of radius r is $4\pi r^3/3$. Our sphere has radius $r = 0.025$ m and hence a volume of 65.4×10^{-6} m^3. The mass m_s of the sphere is the product of its volume and its density ρ_a: $m_s = 2.7 \times 65.4 \times 10^{-6} = 0.1767$ kg and its weight is $m_s g = 0.1767 \times 9.81 = 1.734$ N. The uplift from the water displaced by the sphere is similarly $1 \times 65.4 \times 10^{-6} \times 9.81 = 0.642$ N. The resulting support force in the wire is the difference between the weight of the sphere and the uplift from the water: 1.09 N.

2. A second aluminium alloy sphere is made in two halves and then stuck together such that it too has a diameter of 50 mm but is now hollow with a wall thickness of 1 mm (Fig. 2.18). What force is now required to suspend this sphere or to anchor it to the bottom of the bowl?

The uplift force from the water in the bowl is unchanged at 0.642 N because the volume of water displaced by the sphere is the same. It is the mass and hence weight of the sphere that has changed. Given that the wall thickness of 1 mm is much smaller than the diameter of the sphere we can calculate the volume of aluminium alloy from the product of surface area and thickness. The surface area of a sphere of radius r is $4\pi r^2$. The volume of metal in a sphere of radius r and wall thickness $t \ll r$ is thus $4\pi r^2 t = 4\pi \times 0.0245^2 \times 0.001 = 7.54 \times 10^{-6}$ m^3. Its mass is then 0.0204 kg and its weight is 0.2 N. The weight of the sphere is thus considerably lower than the Archimedean upthrust and the sphere must in fact be anchored down with a force of 0.44 N.

3. What wall thickness of an aluminium alloy sphere of diameter 50 mm would confer the property of neutral buoyancy in fresh water?

Let us call the outer radius of our hollow sphere r_o and the inner radius r_i. Then the volume of metal is:

$$V = \frac{4}{3}\pi r_o^3 - \frac{4}{3}\pi r_i^3 \qquad (2.16)$$

and neutral buoyancy requires that:

$$\rho_a g \left[\frac{4}{3}\pi r_o^3 - \frac{4}{3}\pi r_i^3 \right] = \rho_w g \frac{4}{3}\pi r_o^3 \qquad (2.17)$$

so that:

$$\left[\frac{r_i}{r_o} \right]^3 = \frac{\rho_a - \rho_w}{\rho_a} \qquad (2.18)$$

We have $\rho_a = 2.7\rho_w$ and $r_o = 0.025$ m so that $r_i = 0.021$ m and the required wall thickness of the hollow sphere is 3.6 mm.

2.8 Total and effective stresses

If we now fill our pool with particles of soil and let things settle down, some of the water will have been displaced but there is enough space between the individual soil particles (we will discover how much in Chapter 3) for the water to maintain a continuous presence. If we place a standpipe in the soil/pool (Fig. 2.19) then the water will still rise to the level of the water surface and the pressure in the water will still be calculated from the depth below this water surface. Suppose for initial simplicity that the soil surface coincides with the water surface as shown in Fig. 2.19.

As a little interjection, Fig. 2.20 shows the conventions that we will use to indicate (a) the surface of the ground, (b) the water level and (c) the level of some underlying rock layer. In the absence of any other comment, these symbols will be assumed to convey these meanings.

Our calculation of the vertical stress in the soil required to maintain the equilibrium of the column of soil extending from depth z up to the free surface is still given by (2.10) (we assume that the soil has uniform density ρ):

$$\sigma_z = \rho g z \qquad (2.19)$$

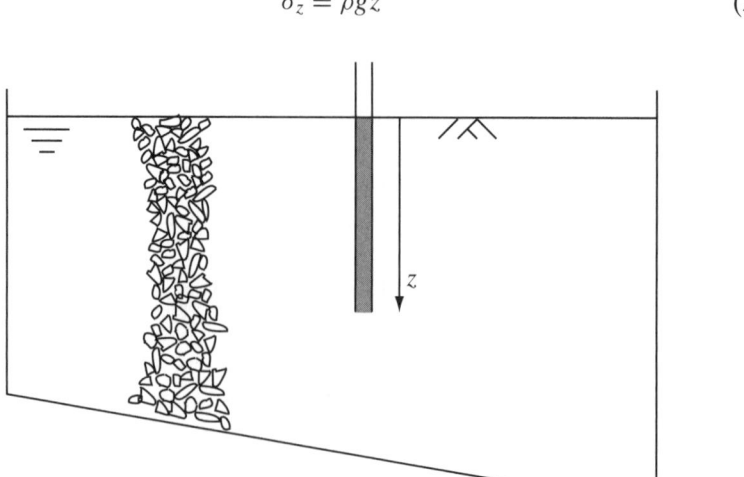

Figure 2.19. Pool filled with soil: water table at surface.

2.8 Total and effective stresses

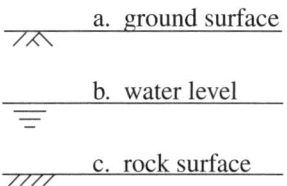

Figure 2.20. Conventions used to indicate (a) the surface of the ground, (b) the water level and (c) the surface of an underlying rock layer.

and the pressure u in the water is given by (2.12). It seems entirely reasonable to suppose that the stress σ_z which is needed to maintain equilibrium – and which is called the *total* stress – is partitioned between the water and the soil. The stress remaining to be carried by the soil particles itself – which is called the *effective* stress, and is traditionally distinguished by adding a mark $'$ – is the difference between the total stress and the water pressure u. The water is present in the pore spaces between the soil particles (see Chapter 3) so we call the water pressure in the soil the pore pressure. Then our partition of the total stress gives us:

$$\sigma'_z = \sigma_z - u \qquad (2.20)$$

This equation is a statement of the *Principle of Effective stress*, which is arguably the most important concept in the mechanics of soils. It is found that, for saturated soils, it is the effective stress that controls all aspects of mechanical response for engineering purposes: the deformability, stiffness and strength are all dependent on effective stress. Though there have been attempts to prove this equation from considerations of the way in which stress is transmitted within soils and between soil particles and through the pore fluid, it is better to accept it as a conjecture for which no particularly serious experimental refutation has been discovered. It must be seen as a completely general statement: as we shall see, there may be all sorts of reasons why there are fluid pressures in the pores of a soil, and the total stress, reflecting the equilibrium of the soil, is likely also to have a more general and more complex source than simply the weight of an overlying column of wet soil. But the Principle of Effective stress states that the effective stress borne by the soil is the difference between the total, equilibrium, stress – whatever its source – and the fluid pressure or pore water pressure – whatever its source.

An illustration of the application of the Principle of Effective stress can be obtained by calculating the vertical effective stresses in the two configurations shown in Fig. 2.21. The element A in Fig. 2.21a is at a depth z beneath the ground surface and the water table is at the ground surface. The water table is the level to which the water rises in the ground. If we dug a hole from the ground surface (or put down a borehole to recover some samples of the soil) we would expect to find water at this level (though for reasons that will become apparent in Chapter 5 it may take a little time for an equilibrium water level to establish itself in the hole).

The density of the soil is ρ and the density of the water is ρ_w. The total vertical stress can be calculated from considerations of equilibrium thinking of the weight

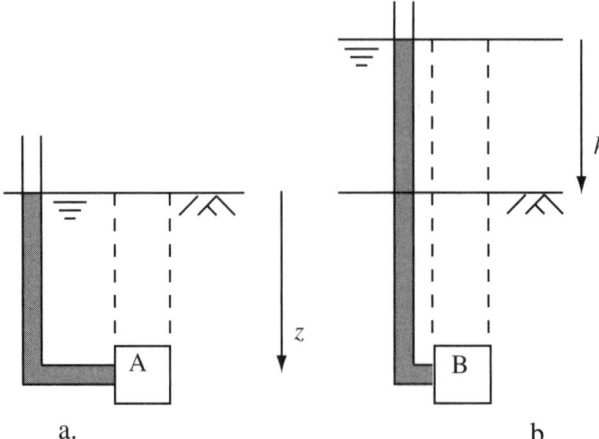

Figure 2.21. (a) Element A at depth z below ground surface, water table at ground surface; (b) element B at depth z below ground surface, beneath lake of depth h.

of the column of material overlying the element A up to the free surface at ground level:

$$\sigma_{zA} = \rho g z \tag{2.21}$$

and the pore water pressure is:

$$u_A = \rho_w g z \tag{2.22}$$

so that the vertical effective stress is:

$$\sigma'_{zA} = \sigma_{zA} - u_A = (\rho - \rho_w)gz \tag{2.23}$$

The element B in Fig. 2.21b is at a depth z beneath the ground surface but the ground is beneath a lake of depth h. The total vertical stress calculated from considerations of equilibrium thinking of the weight of the column of material overlying the element B up to the free surface of the lake is:

$$\sigma_{zB} = \rho g z + \rho_w g h \tag{2.24}$$

and the pore water pressure is:

$$u_B = \rho_w g(z+h) \tag{2.25}$$

so that the vertical effective stress is:

$$\sigma'_{zB} = \sigma_{zB} - u_B = (\rho - \rho_w)gz \tag{2.26}$$

and thus

$$\sigma'_{zA} = \sigma'_{zB} \tag{2.27}$$

The effective stresses are the same even though the total stresses and pore pressures are quite different. Thus, assuming that all other effects are the same (history we will

2.8 Total and effective stresses

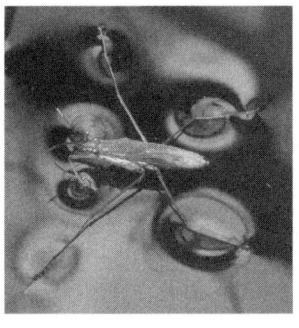
a.

b.

Figure 2.22. (a) Pond skater taking advantage of surface tension to walk on the surface of a pond; (b) pond skater illustrated on a Nagoya man-hole cover.

see also plays its part), the mechanical responses of the two elements of soil should be the same.

There is one other phenomenon associated with the presence of water in the ground that we need to mention. If we have water sitting between the soil particles and if the equilibrium water level seen in a standpipe is not at the ground surface, then there will presumably be an air-water interface somewhere in the soil. Molecules near a liquid surface have an imbalance of attractive forces from neighbouring molecules and, in pulling themselves together, develop a tension in the surface. Work is always required to break or divide a surface – some pond insects are able to walk on the surface of water by exerting contact pressures that are not large enough to break the surface of the pond (Fig. 2.22). If we insert a thin tube through the surface of a liquid (Fig. 2.23a), the liquid will rise until the height of the water in the tube balances the pressure differential across the curved surface of the liquid which results from surface tension: this is *capillary rise*.

We can consider the equilibrium of the little element of the surface shown in Figs 2.23b, c. This is an element of a spherical surface of radius r subtending an angle $\delta\theta$ at the centre of the sphere. The lengths of the sides of the element are thus $\delta\ell = r\delta\theta$. The surface tension forces on each side of the element are $T\delta\ell$ acting tangentially so that the component in the radial direction for our little element is $T\delta\ell \sin \delta\theta/2$. The angle is small so that $\sin \delta\theta/2 \approx \delta\theta/2$. The total radial force from all four edges of the element is then $2T\delta\ell\delta\theta$. The pressure difference between the inside and outside of the surface is Δp and the component of the resulting force in the radial direction is $\Delta p \delta\ell^2$. Equilibrium then requires that:

$$\Delta p \delta\ell^2 = 2T\delta\ell\delta\theta \tag{2.28}$$

or, since $\delta\ell = r\delta\theta$,

$$\Delta p = \frac{2T}{r} \tag{2.29}$$

This is a standard result from analysis of membranes: a spherical membrane of radius r carrying a tensile force T per unit length in all directions (Fig. 2.23b) must

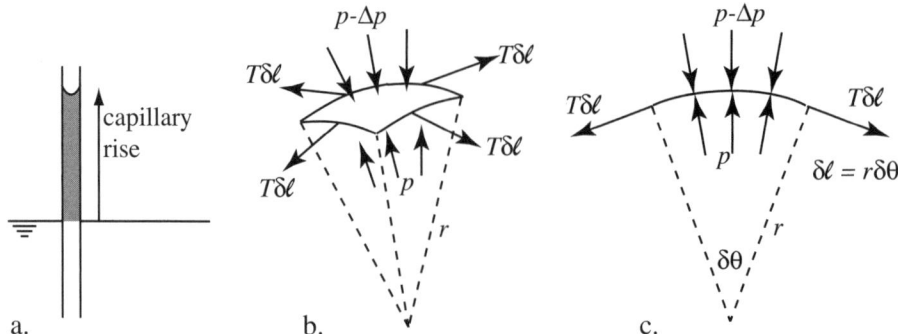

Figure 2.23. Surface tension: (a) capillary rise in small-bore tube; (b) "square" element (side $\delta\ell$) of spherical membrane with in-plane tensile force T per unit length; (c) section through element of membrane.

have a pressure differential Δp across the membrane, as given by (2.29). The balloon in Fig. 2.17 is an example of just such a membrane.

For water, the surface tension T is about 70×10^{-3} N/m at $20°$C so that the radius of tube required to generate a pressure difference across a water surface equal to atmospheric pressure, which is about 100 kPa, would be about 1.4 μm. If our soil contains spaces between the particles which are small enough (see Section 3.5) then capillary effects may be able to develop above the nominal water table and the pore pressure may assume a modestly negative value. For coarse materials, the size of the gaps between the particles will be too large to permit any significant capillary action and the soil can be assumed to be "dry" above the water table.

2.8.1 Worked examples

1. For a site with uniform ground conditions having $\rho = 1.5$ Mg/m^3 and water table at a depth $z_{wt} = 1$ m, calculate the profiles with depth of total and effective vertical stress and pore pressure with and without capillary effects.
   The total vertical stress is given by:
   $$\sigma_z = \rho g z = 1.5 \times 9.81 z = 14.7z \text{ kPa} \tag{2.30}$$
   and the pore water pressure is given by:
   $$u = \rho_w g(z - z_{wt}) = 9.81(z - 1) \text{ kPa} \tag{2.31}$$
   The effective vertical stress is then:
   $$\sigma'_z = (\rho - \rho_w)gz + \rho_w g z_{wt} = 4.9z + 9.81 \text{ kPa} \tag{2.32}$$
   With capillary rise taken into account, these expressions apply throughout the soil. The profiles of stress and pore pressure are shown in Fig. 2.24a and some key values are listed in Table 2.1. Above the water table, the pore water pressure is negative so the Principle of Effective stress (2.20) tells us that the effective stress is greater than the total stress: it has a value of 9.81 kPa at the

2.8 Total and effective stresses

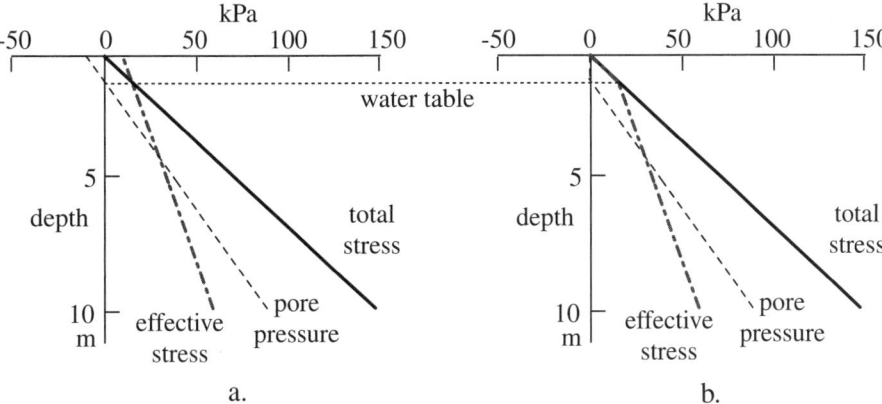

Figure 2.24. Stress profiles: (a) with capillary rise; (b) without capillary rise.

surface. Below the water table the pore pressure is positive and the effective stress is lower than the total stress.

With capillary rise excluded (but assuming that the density of the soil is the same above and below the water table) the expressions for pore pressure (2.31) and effective stress (2.32) do not apply above the water table: the pore pressure is zero and the effective and total stresses are identical. The profiles of stress and pore pressure are shown in Fig. 2.24b and some key values are listed in Table 2.1. The difference between the two cases considered appears only above the water table where the effective stresses are slightly greater if negative pore pressures resulting from capillary effects are included.

2. Figure 2.25 shows a layered soil profile with the water table at the ground surface. Calculate the total vertical stress, pore water pressure and effective vertical stress at the base of each layer: points are labelled A, B, C, D in Fig. 2.25.

This is the same ground profile that was analysed in Section 2.6.1, Example 2 – the only extra feature is the specification of the water table. The total vertical stresses remain as before – equilibrium considerations remain unchanged. The values of σ_z are shown in Table 2.2.

The pore water pressure is calculated directly from the depth below the water table and, with the water table at the ground surface, the pore pressure at depth z is simply $u = \rho_w g z$. Values of u are shown in the table.

Table 2.1. *Stresses in ground for the example shown in Figs 2.24a, b.*

	Capillary rise			No capillary rise		
	σ_z	u	σ'_z	σ_z	u	σ'_z
Location	kPa	kPa	kPa	kPa	kPa	kPa
ground surface	0.0	−9.81	9.81	0.0	0	0
depth 1m (water table)	14.7	0	14.7	14.7	0	14.7
depth 10m	147.1	88.29	58.86	147.1	88.29	58.86

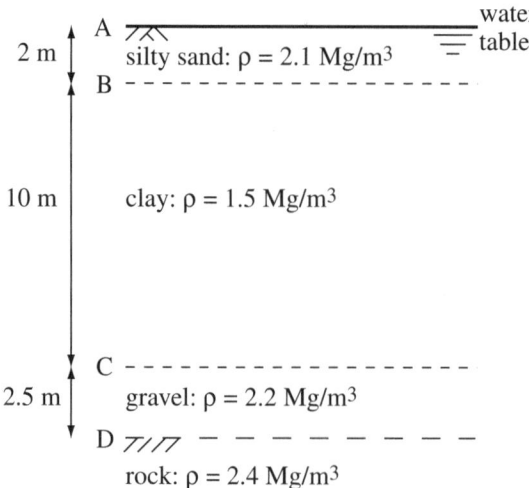

Figure 2.25. Example 2: Profile of layered ground.

Then the effective vertical stress is calculated from the difference between the total vertical stress and the pore water pressure. The effective stress is the stress that remains to be supported by the soil after removing the contribution of the water pressure to the vertical equilibrium of the soil column at any depth. Values of σ'_z are also included in the table.

3. Figure 2.26 shows the same layered soil profile but with the water table at a depth of 1.5 m below the ground surface. Calculate the total vertical stress, pore water pressure and effective vertical stress at the base of each layer: points are labelled A, B, C, D. We will assume that the silty sand layer just beneath the ground surface is able to support capillary suction.

This is the same ground profile that was analysed in Example 2 – the only feature that has changed is the location of the water table. The total vertical stresses remain as before – equilibrium considerations remain unchanged. The values of σ_z are shown in Table 2.2.

The pore water pressure is calculated directly from the depth below the water table and, with the water table at a depth of 1.5 m below the ground surface, the

Table 2.2. Stresses in ground (Figs 2.25 and 2.26).

Location		Water table at surface (Fig. 2.25)			Water table at depth 1.5 m (Fig. 2.26)		
		σ_z kPa	u kPa	σ'_z kPa	σ_z kPa	u kPa	σ'_z kPa
ground surface	A	0.0	0.0	0.0	0.0	−14.7	14.7
base of silty sand	B	41.2	19.6	21.6	41.2	4.9	36.3
base of clay	C	188.3	117.7	70.6	188.3	103.0	85.3
base of gravel	D	242.3	142.2	100.1	242.3	127.5	114.8

2.8 Total and effective stresses

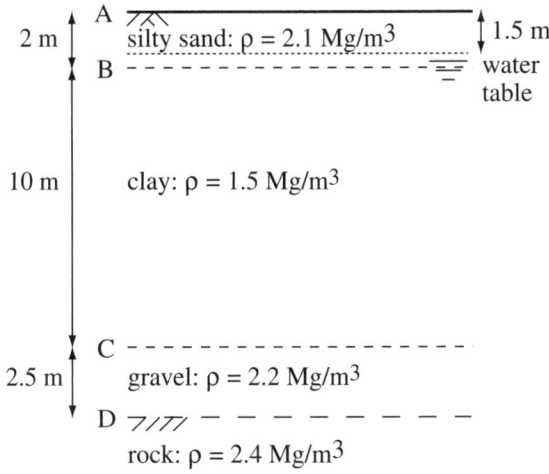

Figure 2.26. Example 3: Profile of layered ground.

pore pressure at depth z is simply $u = \rho_w g(z - 1.5)$. Values of u are shown in the table.

Then the effective vertical stress is calculated from the difference between the total vertical stress and the pore water pressure. Values of σ'_z are also included in the table. Compared with Example 2, the effect of lowering the water table by 1.5 m is to reduce the pore water pressures by 14.7 kPa at all depths – and hence the vertical effective stress is greater at all depths by the same amount.

Thus, the variation of the water table – which is likely to occur as a result of seasonal and climatic effects – has no influence on the total vertical stress but does affect the effective stress, which is the stress that controls the mechanical response of the soil.

4. Figure 2.27 shows a layered soil profile, with the water table at a depth of 1 m below the ground surface. The water table is midway through a layer of sand of thickness 2 m. It is assumed that the sand is sufficiently coarse that it is dry above the water table and there are no capillary effects. However, the density of the sand will be different above and below the water table as shown in the figure. (We will explore the nature of density of soils in Chapter 3.) We want to calculate the total vertical stress, pore water pressure and effective vertical stress at the water table and at the base of each layer: points are labelled A, B, C, D, E.

At the ground surface (A), the vertical total stress $\sigma_z = 0$, the pore water pressure $u = 0$ and the effective stress $\sigma'_z = 0$.

At the level of the water table (B), the vertical total stress is calculated from the density of the overlying dry sand $\rho = 1.7$ Mg/m³: $\sigma_z = 1.7 \times 9.81 \times 1 = 16.7$ kPa, the pore water pressure $u = 0$ and the effective stress is equal to the vertical total stress $\sigma'_z = \sigma_z = 16.7$ kPa.

At the base of the sand layer (C), the vertical total stress is calculated from the sum of the effects of the 1 m of dry sand of density 1.7 Mg/m³ and 1 m of

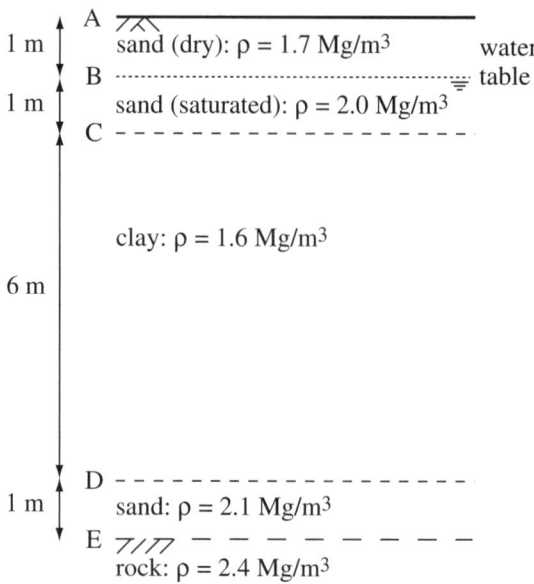

Figure 2.27. Example 4: Profile of layered ground.

saturated sand of density $\rho = 2.0 \, \text{Mg/m}^3$: $\sigma_z = 1.7 \times 9.81 \times 1 + 2.0 \times 9.81 \times 1 = 36.3$ kPa. The pore water pressure at a depth of 1 m below the water table is equivalent to a head of 1 m of water: $u = 9.81$ kPa and the effective stress is the difference: $\sigma'_z = \sigma_z - u = 36.3 - 9.81 = 26.5$ kPa.

At the base of the clay layer (D), the vertical total stress is calculated from the sum of the effects of the 1 m of dry sand of density 1.7 Mg/m³, 1 m of saturated sand of density 2.0 Mg/m³, and 6 m of clay of density $\rho = 1.6 \, \text{Mg/m}^3$: $\sigma_z = 1.7 \times 9.81 \times 1 + 2.0 \times 9.81 \times 1 + 1.6 \times 9.81 \times 6 = 130.5$ kPa. The pore water pressure at a depth of 7 m below the water table is equivalent to a head of 7 m of water: $u = 7 \times 9.81 = 68.7$ kPa and the effective stress is the difference: $\sigma'_z = \sigma_z - u = 130.5 - 68.7 = 61.8$ kPa.

At the base of the lower sand layer (E), the vertical total stress is calculated from the sum of the effects of the 1 m of dry sand, 1 m of saturated sand, 6 m of clay, and 1 m of sand of density $\rho = 2.1 \, \text{Mg/m}^3$: $\sigma_z = 1.7 \times 9.81 \times 1 + 2.0 \times 9.81 \times 1 + 1.6 \times 9.81 \times 6 + 2.1 \times 9.81 \times 1 = 151.1$ kPa. The pore water pressure at a

Table 2.3. Stresses in ground (Fig. 2.27).

Location		σ_z kPa	u kPa	σ'_z kPa
ground surface	A	0.0	0.0	0.0
water table	B	16.7	0.0	16.7
base of sand	C	36.3	9.8	26.5
base of clay	D	130.5	68.7	61.8
base of sand	E	151.1	78.5	72.6

depth of 8 m below the water table is equivalent to a head of 8 m of water: $u = 8 \times 9.81 = 78.5$ kPa and the effective stress is the difference: $\sigma'_z = \sigma_z - u = 151.1 - 78.5 = 72.6$ kPa.

These values are shown in Table 2.3.

2.9 Summary

Here is a concise list of the key messages from this chapter, which are also encapsulated in the mind map (Fig. 2.28).

1. Newton's First Law requires the forces acting on stationary objects to be in equilibrium.
2. Weight is the effect on mass of a local gravitational field. Weight is a force. If mass is measured in kilograms (kg) then force is measured in newtons (N). A mass of 1 kg at the surface of the Earth weighs about 10 N.
3. Stress is an indication of the areal intensity of force and is measured in pascals (Pa): 1 Pa = 1 N/m². Typically, stresses in soils will be quoted in kilopascals (kPa) and compression stresses will be taken as positive.
4. The vertical total stress at a particular depth in the ground can be calculated from the thickness and density of the layers making up the soil column above that depth.
5. At a point in a fluid at rest, the pressure is the same in all directions and is calculated from the depth below the free fluid (water) surface.
6. The total stress in the ground calculated from equilibrium is divided between the water pressure in the pores of the soil and the effective stress which is carried by the soil itself.
7. The Principle of Effective stress is fundamental to understanding of soil behaviour: mechanical properties of soils are controlled by the effective stresses. Effective stress is the difference between total stress and pore water pressure.

2.10 Exercises: Profiles of total stress, effective stress, pore pressure

1. A ground profile consists of 3 m of silty sand of density $\rho = 2.0$ Mg/m³ overlying 12 m of clay with $\rho = 1.6$ Mg/m³ which is underlain by 5 m of gravel with $\rho = 2.3$ Mg/m³ and then rock. Calculate the total vertical stress at depths 0, 3, 15 and 20 m.
2. The water table for the ground profile described in Question 1 is at the ground surface. Calculate the total vertical stress, pore water pressure and effective vertical stress at depths 0, 3, 15 and 20 m.
3. The water table for the ground profile in Question 1 is at depth 1.5 m below the ground surface. Assume that the density of the silty sand is the same above and below the water table. Calculate the total vertical stress, pore water pressure and effective vertical stress at depths 0, 3, 15 and 20 m.
4. The ground profile at a certain site consists of 3 m of coarse sand, overlying 4 m of fine sand, which overlies 2 m of sandy gravel and then rock. The water table

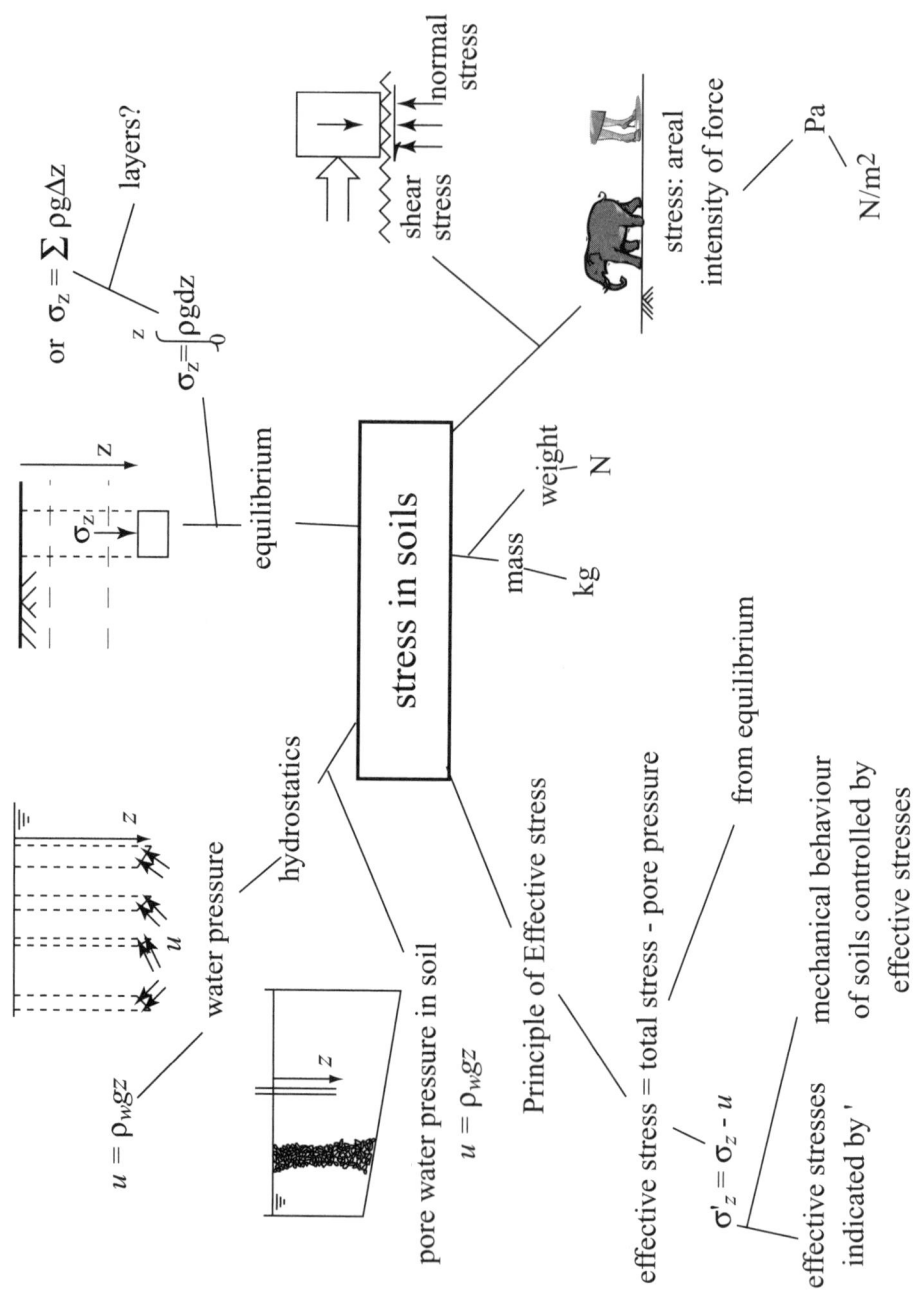

Figure 2.28. Mind map: stress in soils.

2.10 Exercises: Profiles of total stress, effective stress, pore pressure

is at a depth of 3 m. The densities of the three layers are: coarse sand 1.8 Mg/m^3, fine sand 2.0 Mg/m^3, sandy gravel 1.9 Mg/m^3. Calculate the total vertical stress, pore water pressure and effective vertical stress at depths 3, 7 and 9 m.

5. At the site in Question 2.10 the water table rises to a depth of 1.5 m below the ground surface. The density of the coarse sand below the water table is 2.1 Mg/m^3. Calculate the total vertical stress, pore water pressure and effective vertical stress at depths 3, 7 and 9 m, and hence calculate the change in effective stress at each of these depths.

6. At a coastal site the ground conditions consist of 8 m of silty sand, overlying 2 m of gravel and then rock. The density of the silty sand is 1.7 Mg/m^3 above the water table and 1.9 Mg/m^3 below the water table. The density of the gravel is 2.1 Mg/m^3. The water table varies regularly between 1 m and 4 m below the ground surface. What are the maximum and minimum vertical effective stresses at a depth of 4 m?

3 Density

3.1 Introduction

In Chapter 2 we calculated profiles of vertical stress in the ground on the assumption that we knew the value of the density of each of the several layers of soil. We also noted that soils consist of individual particles packed together in such a way that there will generally be spaces between them – voids – which may contain air or water or a combination of air and water (or other fluid). Knowing the relative proportions of space occupied by solid and liquid and gas, and knowing the densities of the individual components, we can estimate the density of the overall mixture that is the soil.

Density is obviously essential for the calculation of stress, but the mechanical behaviour of soils is also strongly influenced by the way in which the soil particles are packed together. It seems intuitively obvious that the greater the proportion of the volume of a material that is occupied by "nothing" the lower will be the resistance of that material to imposition of loads, so we will need to explore density and packing in parallel. It is unfortunate that soil mechanics has been endowed with a plethora of different ways of describing aspects of the distribution of materials within the mixture – the effectiveness of the packing of the particles – and, although really only two or three of these are necessary (and indeed sufficient) for the presentation and understanding of the response of soils, some familiarity is required with all of them. Porous materials obviously occur in other contexts – sintered metals and foams are just two examples – and it is probably inevitable that different disciplines have produced different variables to characterise the nature of the relationship between solid and pore space. Such differences help to preserve the hermetic mystery of the mechanics of soils.

3.2 Units

The *density* of a material is its mass per unit volume, and we will measure it typically in kilograms per cubic metre (kg/m^3). The density of water is $\rho_w = 1000$ kg/m^3 and we can save ourselves some zeroes (or powers of 10) by working in terms of megagrams per cubic metre (Mg/m^3) so that the density of water becomes $\rho_w = 1$ Mg/m^3.

The density of a material is an intrinsic quantity that is not dependent on the gravitational field in which the material finds itself: the density of a chunk of metal (for example) will be the same at the surface of the Earth, on the Moon, or even in a microgravity environment in outer space. However, the calculation of stress in Chapter 2 invariably grouped density with the acceleration due to gravity g to obtain the unit weight, γ:

$$\gamma = \rho g \qquad (3.1)$$

The unit weight of a material thus does vary according to the local gravitational acceleration and will certainly therefore be different on the Earth and on the Moon even if the density of the material (or its constituent parts) remains the same. The unit weight is measured in units of force per unit volume, typically kilonewtons per cubic metre, kN/m³. The unit weight of water is $\gamma_w = \rho_w g = 9.81$ kN/m³ and this is often rounded up to $\gamma_w = 10$ kN/m³ for ease of calculation.

3.3 Descriptions of packing and density

In all aspects of engineering we have to deal with models – appropriate simplifications of reality. Photographs at different scales of soils (Fig. 3.1) show not only that soils consist of solid matter and surrounding voids but also that the solid particles may have very different shapes. In devising ways of describing the packing or arrangement of the particles we begin by trying to define rather simple "first order" variables which are concerned only with the relative volumes of the several constituents and not with more complex information about the shapes of the particles or the nature of their contacts. Thus, if we can see Fig. 3.2a as a schematic picture of the arrangement of solid particles (shown black) with voids which are partially filled with liquid (shown shaded) and partially filled with gas (shown as white bubbles), we can lump together the separate components as shown in Fig. 3.2b and start to analyse the resulting model.

Figure 3.2b shows, on the one hand, the masses of gas (0), liquid (which is usually water) M_w, and solid M_s and, on the other hand, the volumes of gas (which is usually air but might be some other gas such as methane), V_g, liquid, V_w, and solid, V_s and the combined volume of voids $V_v = V_g + V_w$. We will suppose that we know the densities of the water, ρ_w, which is 1 Mg/m³ for fresh water, but somewhat greater for sea water, depending on the salt content (see Section 2.7); and of the soil mineral, ρ_s. The ratio of densities of soil mineral and water is called the specific gravity of the soil mineral, G_s:

$$G_s = \rho_s/\rho_w \qquad (3.2)$$

Some values for specific gravity G_s of minerals which are found in soil particles are given in Table 3.1. Almost all the values of specific gravity lie between 2.5 and 3. Unless there are special circumstances, a good initial guess for the specific gravity of soil particles will be around 2.7: diamond and corundum are included to give an indication that there are some naturally occurring denser minerals. A number

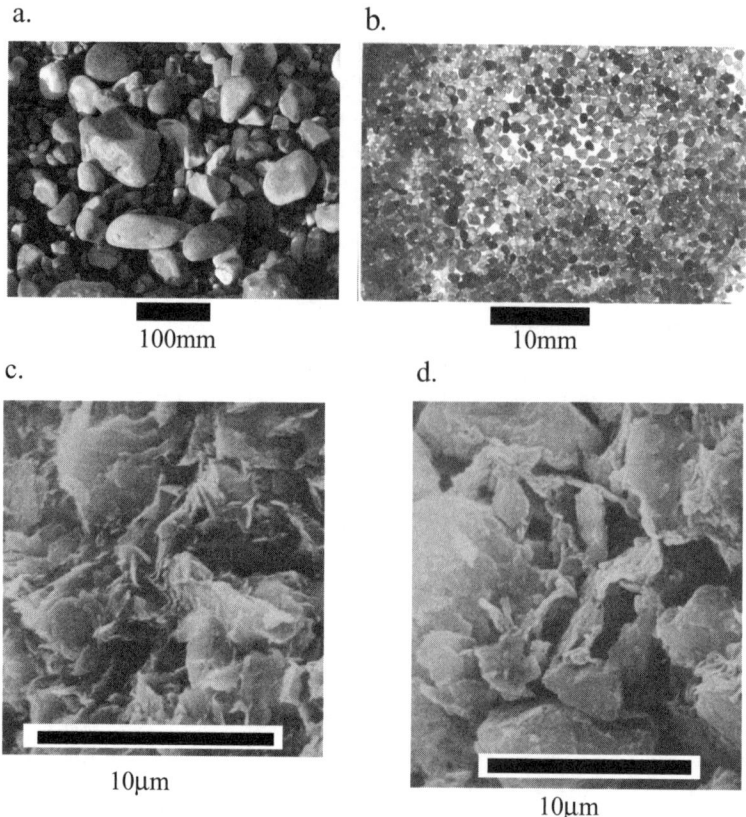

Figure 3.1. (a) Beach shingle; (b) sand; (c) scanning electron micrograph of Weald clay from south-east England (picture provided by A. Balodis); (d) scanning electron micrograph of Drammen clay from Norway (picture provided by A. Balodis).

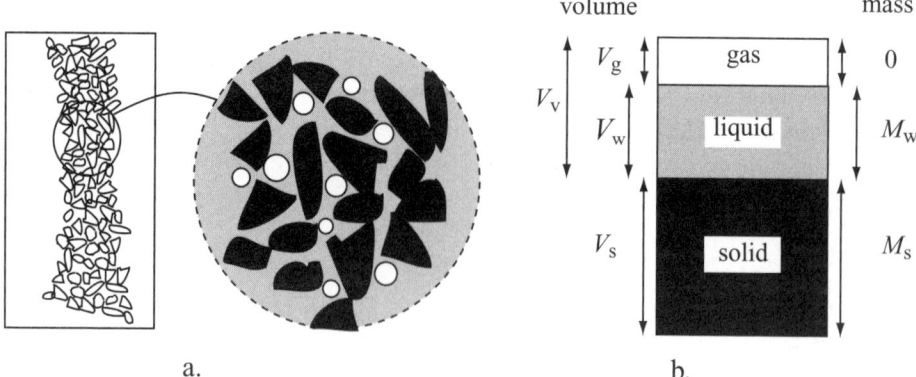

Figure 3.2. (a) Soil particles and surrounding voids partially filled with water; (b) rearrangement of solid, liquid and gas into separate constituent components.

3.3 Descriptions of packing and density

Table 3.1. *Specific gravities of soil minerals.*

Mineral	G_s	Hardness
dolomite	2.85	3.5–4
hornblende	2.9–3.4	5–6
K-feldspar (orthoclase)	2.53–2.56	6
plagioclase feldspar (oligoclase)	2.64–2.68	6–6.5
quartz	2.63	7
corundum	4.05	9
diamond	3.52	10
Clay minerals		
kaolinite	2.6	1.5–2
halloysite	2.6	1–2
illite	2.75	1–2
montmorillonite	2.3–3.0	1–2

of clay minerals are also listed in Table 3.1: these form the rather two-dimensional sheets that can be seen in Figs 3.1c, d but the specific gravity is much the same as that of sand particles or the shingle shown in Figs 3.1a, b.

There are some obvious relationships between the masses and volumes shown in Fig. 3.2b. The volume and mass of the water are linked through the density of the water, and the volume and mass of the solid material are similarly linked through the density of the soil mineral:

$$\rho_w = M_w/V_w \tag{3.3}$$

and:

$$\rho_s = M_s/V_s = G_s \rho_w \tag{3.4}$$

3.3.1 Volumetric ratios

Before we start finding expressions for the actual density of the soil as a combination of solid, liquid and gas we need to define some ratios which can then be incorporated into other definitions. Probably the most important ratio is some description of the closeness of packing of the solid particles: some function of the relative volumes of solid and void. The void ratio, e, is the ratio of the volume of voids (whether this volume is occupied by liquid or gas) and the volume of solid:

$$e = \frac{\text{volume of voids}}{\text{volume of solid}} = \frac{V_v}{V_s} = \frac{V_w + V_g}{V_s} \tag{3.5}$$

In addition to void ratio, there are several alternative ways of expressing the ratios of volumes of void to solid. The specific volume, v, represents the volume occupied by unit volume of solid:

$$v = \frac{\text{total volume}}{\text{volume of solid}} = \frac{V_v + V_s}{V_s} = 1 + e \tag{3.6}$$

In other contexts, where a partitioning approach is to be used to estimate the properties of a mixture, it is common to make use of the solid volume fraction, f_{vol}:

$$f_{vol} = \frac{\text{volume of solid}}{\text{total volume}} = \frac{V_s}{V_v + V_s} = \frac{1}{v} \quad (3.7)$$

Whereas this volume fraction is rarely used in description of soils, it will reappear shortly as a ratio of densities. A more frequently used variable is porosity, n (which is often expressed as a percentage):

$$n = \frac{\text{volume of voids}}{\text{total volume}} = \frac{V_v}{V_v + V_s} = \frac{e}{1+e} = \frac{v-1}{v} \quad (3.8)$$

Finally, we must have some way of recording the fact that the voids are not full of liquid but actually contain a mixture of liquid and gas. The degree of saturation, S_r, describes the proportion of the void space that is filled with water (and is often expressed as a percentage):

$$S_r = \frac{\text{volume of liquid}}{\text{volume of voids}} = \frac{V_w}{V_v} = \frac{V_w}{V_w + V_g} \quad (3.9)$$

But even here there is an alternative variable: the air void ratio, a, which confusingly is a ratio of volume of gas (air) to total volume (and is often expressed as a percentage):

$$a = \frac{\text{volume of gas}}{\text{total volume}} = \frac{V_g}{V_v + V_s} = \frac{(1 - S_r)e}{1+e} \quad (3.10)$$

3.3.2 Water content

Volumes are not easy to measure – especially when the volumes concerned have the irregular boundaries of soil particles. Masses are easier to measure accurately. The most frequently used mass ratio is the water content, w. This is simply the ratio of the mass of water to the mass of dry soil. It can be measured by weighing a sample of soil before and after drying it in an oven. The difference in the masses is the mass of water in the original sample. Thus, water content is given by:

$$w = \frac{\text{mass of water}}{\text{mass of dry soil}} = \frac{M_w}{M_s} \quad (3.11)$$

Water content can be related to the volumetric ratios we have already presented:

$$w = \frac{M_w}{M_s} = \frac{\rho_w V_w}{\rho_s V_s} = \frac{1}{G_s} \frac{S_r e}{1} \quad (3.12)$$

3.3.3 Densities

Having defined these volume ratios, we can next find expressions for density. There are three densities to be defined: the density of the soil in its present, probably unsaturated, state which is the bulk density, ρ; the density of the soil in its current volumetric configuration but with the voids entirely filled with water, the saturated

3.3 Descriptions of packing and density

density, ρ_{sat}; and the density of the soil in its current volumetric configuration but with the voids dried out and filled only with air, the dry density, ρ_d. Each density is, of course, a ratio of mass and volume.

The bulk density is the ratio of the total mass to the total volume:

$$\rho = \frac{\text{mass of solid} + \text{mass of water}}{\text{total volume}} = \frac{M_s + M_w}{V_s + V_v} \tag{3.13}$$

The manipulations required to convert this to an expression involving more fundamental parameters (constituent material densities and volume ratios) are essentially the same whatever eventual quantity we are trying to define. The volume of the voids V_v is:

$$V_v = eV_s \tag{3.14}$$

The mass of water M_w is:

$$M_w = S_r e \rho_w V_s \tag{3.15}$$

The mass of solid M_s is:

$$M_s = \rho_s V_s = G_s \rho_w V_s \tag{3.16}$$

Merging all these expressions, we find that the bulk density ρ is:

$$\rho = \frac{S_r e + G_s}{1 + e} \rho_w \tag{3.17}$$

Expressions for the saturated and dry densities can then be found by setting the degree of saturation S_r in (3.17) to 1 or 0, respectively – indicating that the voids are either completely full of water or completely full of air. Saturated density ρ_{sat} is:

$$\rho_{sat} = \frac{e + G_s}{1 + e} \rho_w \tag{3.18}$$

and dry density ρ_d is:

$$\rho_d = \frac{G_s}{1 + e} \rho_w = \frac{G_s}{v} \rho_w \tag{3.19}$$

Then by comparison with (3.4) we find that specific volume v is in fact simply the ratio of mineral density to dry density:

$$v = \frac{\rho_s}{\rho_d} \tag{3.20}$$

which should not really be a surprise. As the voids become smaller and smaller, and more and more of the volume is taken up by the soil mineral, so the density of the whole must approach the density of the mineral itself. In fact, it will be helpful in future consideration of mechanical response to recall that specific volume and dry density are inversely related.

3.3.4 Unit weights

Calculation of *in-situ* stresses in Chapter 2 always combined density with gravitational acceleration to calculate the profiles of vertical stress. For each density, we can define a corresponding unit weight. In practice, there are three unit weights that are regularly used: the bulk unit weight $\gamma = \rho g$,[1] the unit weight of water $\gamma_w = \rho_w g$, and a buoyant unit weight $\gamma' = \gamma - \gamma_w$.

The buoyant unit weight naturally emerges below the water table where the gradient of increase of total vertical stress σ_z with depth is:

$$\frac{d\sigma_z}{dz} = \rho g = \gamma \quad (3.21)$$

and the gradient of increase of pore water pressure with depth is:

$$\frac{du}{dz} = \rho_w g = \gamma_w \quad (3.22)$$

We saw in Section 2.8 that the behaviour of soils is controlled by an effective stress which is calculated as the difference between total stress and pore pressure. The total stress guarantees equilibrium and the water pressure takes some of this stress – the soil carries the remainder. We defined effective stress σ'_z using the Principle of Effective stress:

$$\sigma'_z = \sigma_z - u \quad (3.23)$$

If we work in gradients of the several components of stress with depth, we can write:

$$\frac{d\sigma'_z}{dz} = \frac{d\sigma_z}{dz} - \frac{du}{dz} = \gamma - \gamma_w = \gamma' \quad (3.24)$$

thus demonstrating the relevance of the buoyant unit weight.

A word of warning, however. The use of buoyant unit weight to calculate variations of effective vertical stress is limited to situations where there is no flow occurring. We will encounter effects of seepage and flow in Chapter 5. If in doubt, calculate effective stresses from total stresses and pore pressures rather than relying on buoyant unit weight γ' to produce the correct result. The significance of this warning will be appreciated subsequently.

3.3.5 Typical values

It is helpful to have an idea of the typical values of density and unit weight that engineering soils are likely to have. This not only provides some control on the plausibility of quoted values – so that "impossible" values can be identified and eliminated – but also leaves the possibility of making use of reasonable estimates if no actual measurements of density are available.

---

[1] Beneath the water table, the soil will usually be more or less saturated so that the bulk unit weight γ will be the saturated unit weight.

3.4 Measurement of packing

Table 3.2. *Typical values of soil densities.*

Soil description	n	e	w	ρ_{sat} Mg/m^3	ρ Mg/m^3	ρ_d Mg/m^3
sandy gravel (loose)	0.4	0.67		2.05	2.01	1.63
sandy gravel (dense)	0.3	0.43		2.23	2.18	1.9
soft clay		2.2	0.81	1.53	1.5	0.84
stiff clay		0.8	0.3	1.94	1.92	1.5

The specific gravity of soil minerals is typically between 2.6 and 2.8 – in the absence of other information we can take a value $G_s = 2.7$. A typical sand might have void ratios which cover a range from 0.4 to 1. From (3.19), dry densities should then fall into the range 1.35-1.9 Mg/m^3 and saturated densities 1.85-2.2 Mg/m^3. Clays are able to survive with much larger void ratios, up to 2 or more, for example (note the very open structures apparent in Figs 3.1c, d), leading to dry densities as low as 0.9 Mg/m^3 and saturated densities around 1.5 Mg/m^3. Some typical values are given in Tables 3.2 and 3.3.

3.4 Measurement of packing

Having defined various measures of density or packing, we need to have some way of determining them. Density is mass divided by volume: measurement of masses is relatively straightforward; measurement of volumes of irregular objects such as chunks of soil or collections of soil particles is not so straightforward.

The easiest quantity to measure is water content, this is simply a ratio of masses (3.11). We use a weighing dish or bottle: weigh the dish on its own, M_1, then weigh it with the moist soil, M_2, then dry the soil on the dish in an oven to drive off the water and, finally, weigh the dish with the dry soil, M_3. The mass of dry soil is $M_3 - M_1$, the mass of water in the sample of soil is $M_2 - M_3$, and the water content w is then:

$$w = \frac{M_2 - M_3}{M_3 - M_1} \quad (3.25)$$

This tells us nothing about whether or not the soil is saturated but, if we were to suppose that the soil were indeed completely saturated ($S_r = 1$), then we could

Table 3.3. *Typical values of soil unit weights.*

Soil description	n	e	w	γ_{sat} kN/m^3	γ kN/m^3	γ_d kN/m^3	γ' kN/m^3
sandy gravel (loose)	0.4	0.67		20.1	19.7	16.0	10.3
sandy gravel (dense)	0.3	0.43		21.9	21.4	18.6	12.1
soft clay		2.2	0.81	15.0	14.7	8.2	5.2
stiff clay		0.8	0.3	19.0	18.8	14.7	9.2

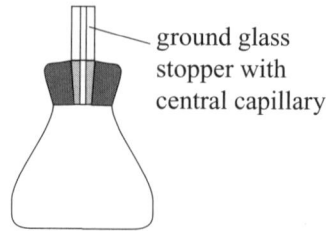

Figure 3.3. Specific gravity bottle, or pycnometer.

calculate the void ratio, e (or any other measure of volumetric packing), from (3.12):

$$e = G_s w \tag{3.26}$$

However, this conversion requires knowledge of the specific gravity of the soil particles, G_s.

Accurate determination of specific gravity requires the use of a specific gravity bottle, or pycnometer (Fig. 3.3). The specific gravity bottle is a carefully made glass bottle with a closely fitting ground glass stopper through which there is a narrow tubular hole. When the bottle is filled with liquid and the stopper inserted, surplus liquid escapes through the hole and can be carefully wiped clear in such a way that the volume of liquid contained in the bottle is accurately the same every time the bottle is used.

Specific gravity is a ratio of densities of a substance under study and pure water (3.2). If the substance is a liquid other than water (for example, sea water or some other saline pore fluid, Section 2.7), then we can determine its specific gravity by weighing the bottle first empty, M_1, then full of water, M_2, and finally full of the liquid, M_3. The volumes of liquid and water are identical, so the specific gravity of the liquid, G_l is:

$$G_l = \frac{M_3 - M_1}{M_2 - M_1} \tag{3.27}$$

To determine the specific gravity of the soil mineral we proceed in a slightly different way. First, we weigh the bottle on its own, M_1, then place some of the dry soil particles in the bottle and weigh it again, M_2. Then we fill the bottle, still containing the soil particles, with water, making sure that absolutely no air is trapped in or around the soil and weigh it again, M_3. Finally, having emptied out the soil and cleaned the bottle, we fill it with water and weigh it, M_4. The mass of the soil particles is evidently $M_2 - M_1$. The mass of the quantity of water which has a volume identical to the volume of the soil particles is then $(M_4 - M_1) - (M_3 - M_2)$, and hence the specific gravity of the soil particles is:

$$G_s = \frac{M_2 - M_1}{M_4 - M_3 + M_2 - M_1} \tag{3.28}$$

To discover the degree of saturation we have no choice but to determine the density of the soil. For this we need to know the volume of the soil sample whose

3.4 Measurement of packing

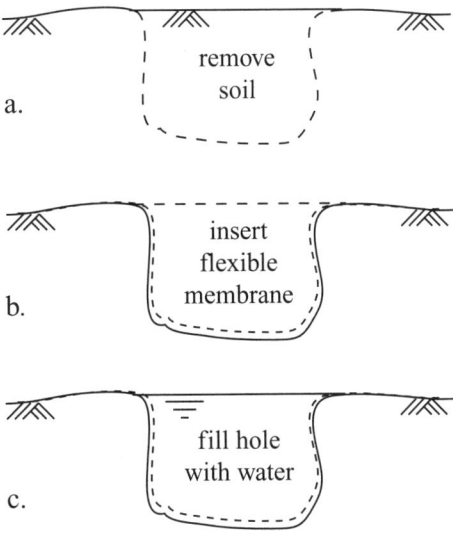

Figure 3.4. Determination of *in-situ* density by filling sample hole with water.

mass we can easily discover. Some soils – much like plasticine,[2] or pottery clay – can be carved into quite precise geometrical forms whose size can be accurately measured. Others – such as sand on the beach – tend to fall apart and form much more irregular shapes. We have to adopt intermediary stratagems. An example is shown in Fig. 3.4.

We have suggested, and we will subsequently reiterate, that the density of packing of soils is important in controlling their mechanical properties: usually the denser the better in terms of strength and stiffness. When man-made soils, or fills, are to be placed as part of a geotechnical structure – a dam (Figs 1.12, 1.13), a road embankment (Fig. 1.6), or behind a retaining wall (Fig. 1.2) or fortification (Fig. 1.1), for example – it is usually necessary to ensure that the soil has the required density specified in the design: payment to the contractor will probably be dependent on meeting this specification. *Compaction* is the name given to the process of improving the density of such fills by using rollers with or without vibration. Inevitably, density comes at a price because it may require a lot of effort and many passes of the roller to increase the density sufficiently. One way of discovering the actual density that has been achieved is to dig a little hole at the surface of the newly placed and compacted fill and carefully collect all the soil (Fig. 3.4a). Then we can place a thin membrane in the hole (Fig. 3.4b) and measure the quantity of water required to fill it (Fig. 3.4c). From the volume of the water we know the volume of the hole, and we can measure the mass of the extracted soil. We can then dry the sample and determine its dry mass, so that we can calculate both the *in-situ* bulk density ρ and

---

[2] Plasticine is used in the United Kingdom as a generic term for a modelling clay, formed from calcium carbonate, petroleum jelly and aliphatic acids, which is popular with children and with animators such as Nick Park for the characters Wallace and Gromit.

the dry density ρ_d of the soil. We can calculate the void ratio or specific volume from the dry density (3.19):

$$e = G_s \frac{\rho_w}{\rho_d} - 1 \qquad (3.29)$$

and then the degree of saturation from the bulk density (3.17):

$$S_r = \frac{(1+e)\rho/\rho_w - G_s}{e} \qquad (3.30)$$

3.4.1 Compaction

As a slight digression, and as an example of the application of some of the measures of volumetric packing and density that have been introduced in this chapter, we can pursue a little further the discussion of *compaction* of soils. It is clearly important for a geotechnical contractor to be confident that the design density will be achieved with a reasonable amount of compaction effort. Laboratory trials are conducted to discover the water content (or moisture content – the terms are interchangeable, and moisture content is more often used in this context) for which the maximum dry density (and hence maximum closeness of particles) can be obtained for a given amount of compaction effort. Recall that dry density correlates inversely with specific volume (3.20). A typical result is shown in Fig. 3.5.

For low moisture contents, the soil is too dry to be compacted easily. As we add more water, so the ease of compaction increases and thus the density attainable for a given compaction effort increases with moisture content. However, there is a curve of full saturation which can be plotted in this diagram and, as this curve is neared, it becomes increasingly difficult to produce any further increase in density. Compaction is concerned with trying to use mechanical impacts (from a moving roller or falling weight) to produce immediate increases in density. We will see how difficult, if not impossible, this can be in saturated fine grained soils in Chapter 6. In fact, there is a tendency for the link between dry density and moisture content to be limited by a line of constant, small, air void ratio (3.10) so that, as the moisture content is increased beyond an optimum value, the density that can be obtained for a given amount of compaction energy actually falls off. The optimum moisture content, giving the maximum dry density (or minimum specific volume), is thus rather important. As the amount of compaction energy increases, so the maximum dry density also increases but, because of the limitation set by the air void ratio line, the optimum water content reduces slightly (Fig. 3.5).

We can plot contours of degree of saturation on the complementary diagrams linking dry density or specific volume and moisture content (shown as solid lines in Fig. 3.5); and we can also plot contours of air void ratio on the same diagrams (shown as dotted lines in Fig. 3.5). Evidently, zero air void ratio ($a = 0$) corresponds exactly with complete saturation ($S_r = 1$). The compaction curve for the particular soil is shown with the heavy solid line: it is becoming more or less parallel with the air void ratio line $a = 10\%$. The heavy dashed line in Fig. 3.5a shows the effect of reducing

3.4 Measurement of packing

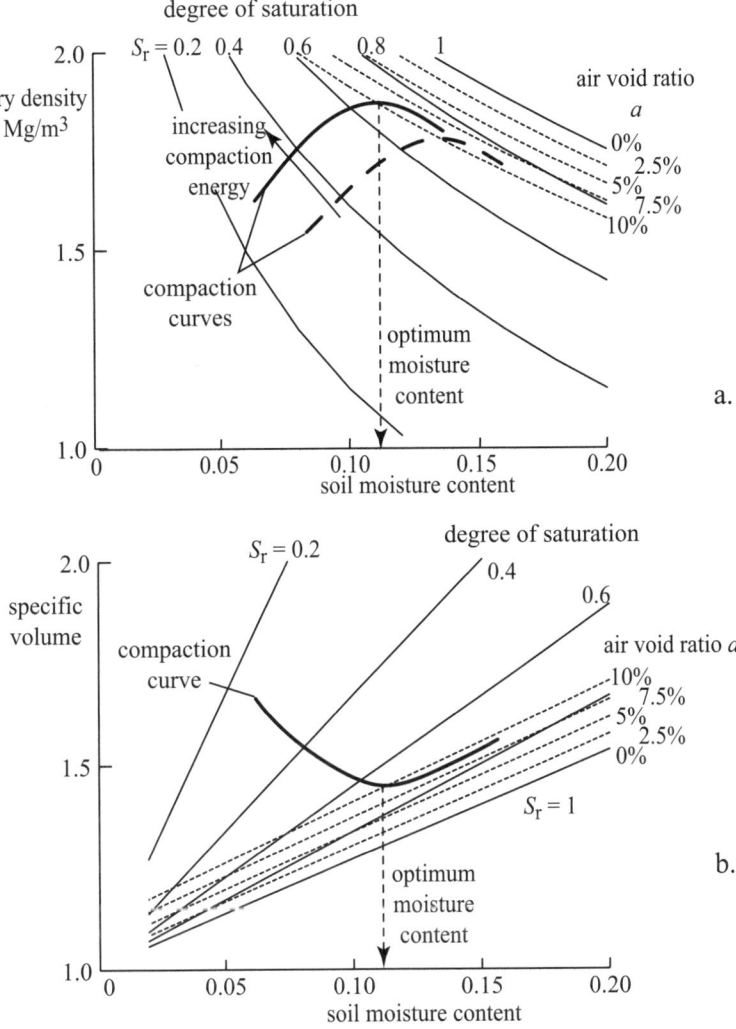

Figure 3.5. Typical compaction curves linking (a) dry density and moisture content and (b) specific volume and moisture content for a given amount of compaction effort.

the amount of compaction effort. The compaction curve is shifted down and to the right so that the maximum density that can be obtained is somewhat reduced and the optimum moisture content is somewhat increased.

Although traditionally compaction information is plotted in terms of dry density and water content, when plotted in terms of specific volume and water content (Fig. 3.5b) the contours of constant degree of saturation and of constant air void ratio become straight lines. Manipulation of the defining equations reveals that:

$$v = 1 + \frac{G_s w}{S_r} \qquad (3.31)$$

which produces a series of radiating lines through the point $v = 1$ and $w = 0$; and:

$$v = \frac{1 + G_s w}{1 - a} \qquad (3.32)$$

which produces a series of lines of slope $G_s/(1-a)$ which for the small values of a plotted in Fig. 3.5b appear almost parallel.

3.5 Soil particles

Even a cursory glance at the pictures of soil particles in Fig. 3.1 should trigger two thoughts: the very wide range of sizes of particles that may make up a soil; and the enormous contrasts in particle shape. Soil particles are formed by a combination of processes of erosion, weathering and transport from parent rocks. The constituents of these rocks are minerals such as quartz, feldspar, mica and hornblende, which are more or less resistant to effects of temperature, humidity and chemical weathering. The specific gravities of many commonly occurring minerals were given in Table 3.1. That table also gives the hardness values of the various minerals on Moh's hardness scale (Table 3.4: fingernails have a hardness of about 2.5). Quartz, a mineral whose chemical composition is silicon dioxide (SiO_2), is a particularly stable and hard mineral which makes up much of the sand and gravel deposits of the world – it may succumb to mechanical effects as the particles are bounced around the surface of the Earth so that particles become progressively more round as the corners are removed (look at the very round shingle particles in Fig. 3.1a) but the chemical composition remains intact. The transport process is clearly important in controlling the nature of the soils that we see around us today.

The four means of transport are by gravity (as a landslide or debris flow), ice, water and air (wind) and, moving through this sequence, the size of "particles" that can be transported becomes progressively smaller. At one extreme, debris flows can include enormous boulders but wind will only lift rather small particles. Ice – in the form of glaciers – smoothes the underlying rocks and is able to carry a great range

Table 3.4. *Moh's hardness scale.*

Hardness	Mineral
0	liquid
1	talc
2	gypsum
3	calcite
4	fluorite
5	apatite
6	orthoclase
7	quartz
8	topaz
9	corundum
10	diamond

3.5 Soil particles

Figure 3.6. Ice movement in glaciers smoothes the underlying rocks and carries a great range of particle sizes (Nigardsbreen, Jostedalen, Norway).

of particle sizes including reasonably large boulders (Fig. 3.6). The size of particle that can be moved by water depends on the speed of flow of the water. As the speed of flow of a river reduces, so the coarser particles will be progressively deposited. The finer particles will be able to remain in suspension for longer – their rate of fall through the water will be slow. A river disemboguing into a lake or sea will still be able to carry the finest particles even past the mouth of the river (Fig. 3.7). The fine particles will slowly fall to the lake floor, forming a layered sediment which may well show some seasonal variations in particle size depending on the flow rates in the tributary rivers.

However, some rock minerals are more susceptible to chemical decomposition and clay minerals such as those seen in Figs 3.1c, d are decomposition products.

Figure 3.7. Island in the Irrawaddy (Myanmar) streaked with silt by receding floodwaters.

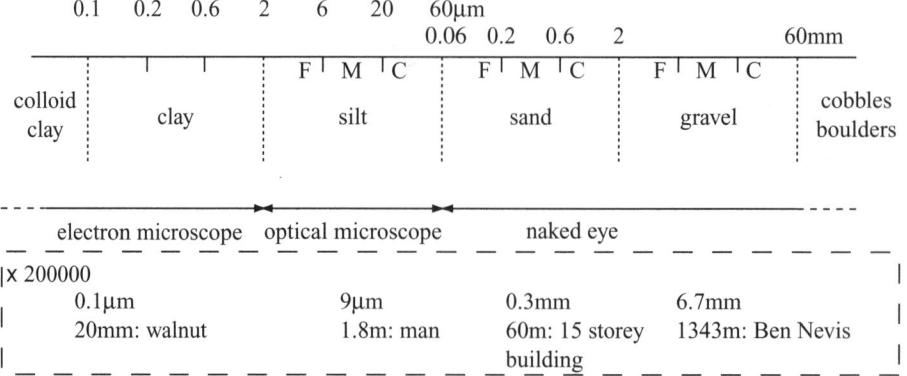

Figure 3.8. Classification of soils by particle size.

Clay minerals appear as packets of alumino-silicate molecules, characterised by large surface area and large ratio of lateral dimensions to thickness. Three of the most common clay minerals – kaolinite, illite and montmorillonite – are decomposition products of feldspars and volcanic ash. Clay minerals have a very low hardness (Table 3.1): hardness value 1-2 implies that they might just about leave a mark on a piece of paper. They are found in packets of molecules of small size which are easily transported by water or ice.

The packets of clay molecules are small: soils are divided into broad categories depending on the size of the particles that can be discovered by standard laboratory tests (Fig. 3.8) – boulders, gravel, sand, silt, clay – with each descriptor covering a range with roughly the same ratio of maximum to minimum sizes (the letters F, M, C for silt, sand and gravel indicate fine, medium and coarse ranges within the overall particle descriptor). Even leaving aside massive boulders and taking a maximum particle size of, say, 10 mm which is in the middle of the gravel range and looking down to 0.1 μm towards the fine end of the clay range, there is a range of sizes of 100,000 (10^5). Multiplying the sizes by 200,000, the fine clay particle becomes 20 mm – a walnut – a 9 μm medium silt particle becomes 1.8 m – a man – a 0.3 mm medium sand particle becomes 60 m – the height of a 15 storey building or the height to the axis of a modest wind turbine – and a 6.7 mm gravel particle becomes 1343 m, which is the height of Ben Nevis, the highest point in the British Isles.

Typical distributions of particle sizes are shown in Fig. 3.9, which uses a logarithmic scale of particle size on the horizontal axis and plots the percentage by mass finer than any given particle size. Techniques that are used to determine such particle size distributions will be described in Section 3.6. The particle size distribution for any particular soil can be characterised by the ratio of the particle size (d_{60}) for which 60% of the material is finer and the particle size (d_{10}) for which 10% of the material is finer. The coefficient of uniformity, C_u, is defined as (Fig. 3.10):

$$C_u = d_{60}/d_{10} \tag{3.33}$$

3.5 Soil particles

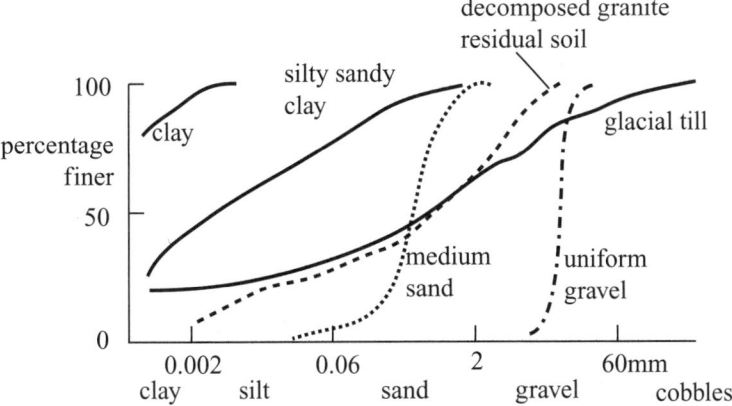

Figure 3.9. Typical particle size distributions.

Soils with low values of C_u are described as poorly graded: a single sized material would have $C_u = 1$. Soils with larger values of C_u are described as well graded. The decomposed granite residual soil in Fig. 3.10 has a coefficient of uniformity of about $C_u \approx 380$.

In hot, humid, climates granite decomposes *in situ* eventually becoming a residual soil. Weathering progresses from planes of weakness within the rock so that at an intermediate stage of decomposition the "soil" contains everything from boulders to clay minerals: kaolinite is a decomposition product of granite (Fig. 3.11). Figure 3.9 shows the very wide range of particle sizes that such natural residual soils and glacial tills can contain. What such a diagram does not indicate is the nature of the particles at the smaller size: very small particles may indeed be clay molecules or packets of clay molecules but they may also be very finely ground rock particles – rock flour – which will have very different properties of mechanical interaction. The scanning electron micrographs of natural clays in Figs 3.1c, d show just such combinations of clay mineral molecules and small rock particles.

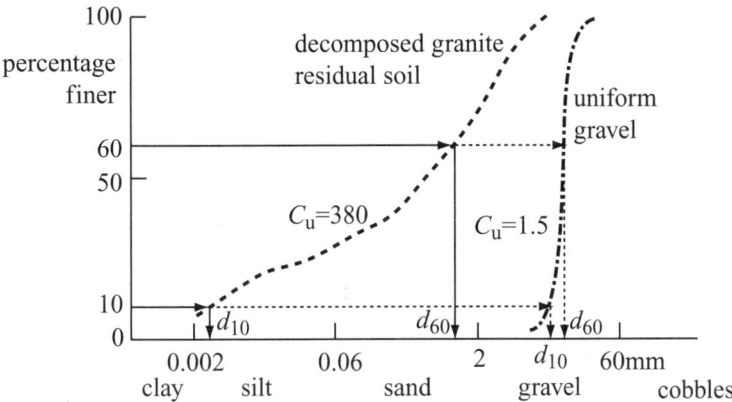

Figure 3.10. Particle size distribution: definition of coefficient of uniformity $C_u = d_{60}/d_{10}$.

10 metres (approximately)

Figure 3.11. Excavation through *in-situ* weathered granite (Sha Tin, New Territories, Hong Kong).

Looking back at the photographs of soil particles (Fig. 3.1) we see that the sand grains and shingle have a compact shape with a low specific surface, the ratio of surface area to volume. A sphere is the solid shape which has the lowest specific surface. The interaction between such particles is broadly mechanical through transmission of forces at the contacts with neighbouring grains. On the other hand, the clay has a high specific surface and much smaller particles. As clay molecules assemble, the distribution of charge is not uniform and there is variation of the electrostatic charge over the surface of the molecules: the faces of the molecules are positively charged and the edges negatively charged. Groups of clay molecules are formed in which positively charged surfaces attract negatively charged particles to form more or less stable packets. The electrostatic forces between adjacent particles are the most significant contributor to the transmission of forces through the particulate assembly for a clay. The presence of such electrostatic forces leads to rather open structures with high void ratios and consequent low densities (Figs 3.1c, d and Tables 3.2, 3.3).

3.6 Laboratory exercise: particle size distribution and other classification tests

3.6.1 Sieving

The distributions of particle sizes shown in Fig. 3.9 treat the soil particles as though they were equivalent spheres – this seems to be a good starting point in our soil

3.6 Laboratory exercise: particle size distribution and other classification tests

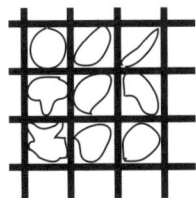

Figure 3.12. Particle sizes determined by sieving.

classification model even if we know that the particles may actually not be particularly spherical. An obvious way to discover the proportions of different sized particles is by sieving, using standard sieves with different mesh sizes. The soil is poured into the top of a stack of sieves, with the coarsest sieve at the top. The stack is shaken and the quantity of material retained on each sieve is measured. Figure 3.12 illustrates the way in which particles of very different shapes will be able to pass through the same square mesh opening and be recorded as below that particular equivalent diameter: this is a first order model for which this approximate equivalent spherical estimate will suffice.

3.6.2 Sedimentation

For a soil with finer particles, sieving is no longer feasible. It would not be possible to construct a sieve with a size of opening small enough to retain particles with a typical dimension of a few microns. An alternative procedure is required. Stokes' Law describes the terminal velocity of spheres falling through a viscous fluid (Fig. 3.13). We can approach Stokes' Law through a consideration of the resistance to the movement of a spherical object through a viscous liquid using the ideas of dimensional analysis.[3]

The resistance force or drag on an object can be expected to depend on the viscosity of the fluid which is trying to resist the motion of the object. We will meet viscosity in Chapter 5 when we look at the way in which water flows through the restricted passages provided by natural soils. It is an indication of the shear stress generated when there is a velocity gradient in the fluid: the faster-flowing fluid tries to pull the slower fluid along with it. The material constant of proportionality linking shear stress and velocity gradient is the viscosity, η. Viscosity has dimensions of mass/(length×time), so it could be expressed in units of kilograms per metre per second. However, understanding its origin as a description of the shear stress generated by a velocity gradient, it is more appropriate to see it in units of pascal-seconds. Viscosity is also quoted in units of centipoise: 1 centipoise = 1 mPa-s. The viscosity of water at room temperature is about 1 mPa-s, 1 millipascal-second (or 1 centipoise). The viscosity of water is quite sensitive to temperature, as the values quoted in Table 3.5 indicate.

---

[3] See, for example, Palmer, A.C. (2008) *Dimensional analysis and intelligent experimentation*, World Scientific Publishing Company, Singapore.

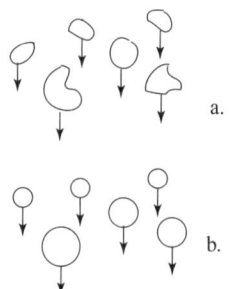

Figure 3.13. Particle sizes determined by sedimentation: (a) actual soil particles; (b) equivalent spherical particles.

Viscosity is relevant in calculating the resistance force on the object because the object partly blocks the flow and forces the fluid to move aside and hence move faster in the vicinity of the object. The resulting velocity gradient then implies the generation of shear stresses in the viscous fluid. Figure 3.14 illustrates the way in which the fluid flows past a stationary spherical object (a problem exactly complementary to the one of a spherical object falling through the "stationary" fluid). We know intuitively that it is much more difficult to pull an object through a low viscosity fluid such as water than through a much more viscous fluid such as treacle or molasses.

At the most basic level, whenever we perform an analysis or a calculation we need to ensure dimensional consistency of the various terms or factors. Dimensional analysis goes further and indicates that we can simplify the analysis of a problem if we arrange the several controlling parameters or variables into dimensionless groups, and it also tells us how many such dimensionless groups there must be. Dimensional analysis does not reveal the form of the relationships between the dimensionless groups, but correct use of the dimensionless groups makes parametric studies more efficient by revealing which variables are truly independent, and also forms the basis for extrapolating from one scale of observation to another.

In this case, the number of parameters or material properties is rather limited. We expect the drag F to increase with the size of object (the radius r of our spherical particle), with the speed V of the object relative to the fluid, and with the viscosity η

Table 3.5. *Viscosity of water.*

Temperature °C	Viscosity mPa-s
10	1.304
15	1.127
20	1.002
25	0.891
30	0.798

3.6 Laboratory exercise: particle size distribution and other classification tests

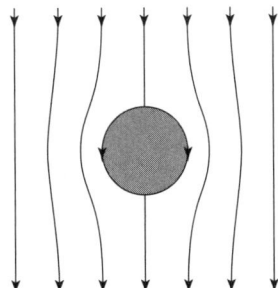

Figure 3.14. Fluid flow around a spherical object.

of the fluid. In fact, the only dimensionless grouping that we can form with the four controlling parameters is $F/rV\eta$. We conclude that this dimensionless group should be a constant independent of the values of the individual quantities in particular applications. Dimensional analysis does not allow us to work out what the constant is; for this some analytical procedure is required. Stokes demonstrated that:

$$\frac{F}{rV\eta} = 6\pi \qquad (3.34)$$

Now, if we have a spherical object in free fall in a viscous fluid, having reached its terminal velocity, then it is falling without acceleration and the viscous drag must exactly balance the buoyant weight of the object (the buoyant weight allows for the Archimedes uplift on any immersed body, Section 2.7). So, if the spherical object has density ρ_s and radius r and is falling at velocity V through a liquid of density ρ_w and viscosity η, then the buoyant weight of the spherical particle is exactly balanced by the drag force:

$$6\pi r V \eta = \frac{4}{3}\pi r^3 (\rho_s - \rho_w) g \qquad (3.35)$$

and thus the terminal velocity of this spherical object is:

$$V = \frac{2gr^2}{9\eta}(\rho_s - \rho_w) \qquad (3.36)$$

These equations apply provided the concentration of settling spherical particles is low so that one particle does not interact significantly with any adjacent particles.

In our laboratory experiment, we need to work with an initially well-mixed dilute suspension of soil particles in a tube of water – the dilute suspension is required to ensure that we can reasonably think of the soil particles as individual, non-interacting equivalent spheres. We add a deflocculating agent to the water to ensure that surface electrostatic forces do not encourage the finer particles to congregate in larger flocs, which would frustrate the attempt to discover the true size distribution of the soil particles. Having shaken and stirred the suspension, we stand the tube carefully and wait. The particles start to settle. We assume that all particles reach their steady, terminal velocity (3.36) immediately.

We now take small samples of the liquid together with suspended particles from the same depth z_r in the tube at different, carefully selected, times (t_1, t_2, t_3, t_4 in

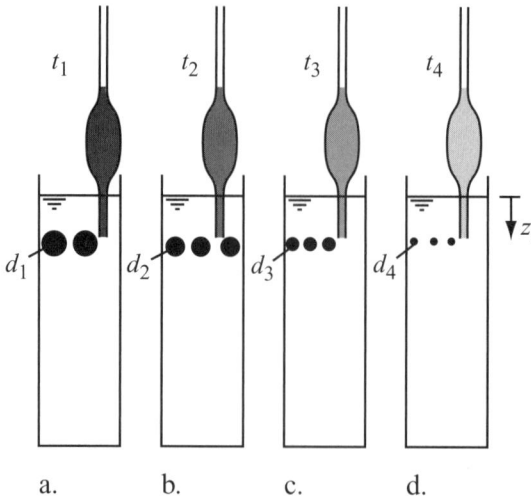

Figure 3.15. Sedimentation experiment for determination of distribution of finer particles.

Fig. 3.15) and then dry the samples in an oven to determine the quantity of solid material that has been extracted. Comparison of these quantities of solid material enables us to compute a number of points on our particle size distribution curve for the finer sizes of particle. At time t_1, particles of size d_1 are just disappearing below the depth z_r but the sample extracted will contain a representative sample of all particle sizes less than and equal to d_1. We can find the link between values of particle size d and time t from (3.36). We know that the terminal velocity of equivalent spheres of radius $r = d/2$ is $V = z_r/t$. Thus:

$$d = \sqrt{\frac{z_r}{t} \frac{18\eta}{g(\rho_s - \rho_w)}} \quad (3.37)$$

We continue to take further samples from the same depth. At time t_2, particles of size $d_2 < d_1$ are just disappearing below depth z_r and the sample will contain all particle sizes less than and equal to d_2. Similarly, at time t_3 the sample contains all sizes less than and equal to d_3, and at time t_4 all sizes less than and equal to size d_4. (We also need to take a sample of the liquid without any suspended soil from another tube in order that we may know the quantity of the deflocculating agent that will be present in each sample.)

We can choose the sampling times to correspond with helpful boundaries on the particle size chart (Figs 3.8, 3.9) so that d_1, d_2, and so on are, say, 20 μm, 6 μm, 2 μm, 0.6 μm and 0.2 μm giving us, with a little calculation, the proportions of medium silt, fine silt, and various clay fractions. For example, let us suppose that the density of the soil mineral is $\rho_s = 2.65$ Mg/m^3 and the water through which the soil is settling has density 1 Mg/m^3. The sedimentation tube is kept in a heated tank with constant temperature 25°C. At this temperature, water has a viscosity of 0.891 mPa-s (Table 3.5). Then we can calculate the time at which particles of equivalent

3.6 Laboratory exercise: particle size distribution and other classification tests

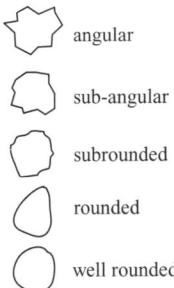

Figure 3.16. Particle shapes.

spherical diameter 6 μm (radius $r = 3$ μm) will have settled below a depth of 100 mm below the surface of the water. This is the size of particles at the boundary between medium silt and fine silt, and thus at this time all particles of medium silt size and above will have settled past our sampling depth:

$$t = \frac{z_r}{V} = \frac{9\eta z_r}{2gr^2(\rho_s - \rho_w)}$$
$$= \frac{9 \times 0.891 \times 10^{-3} \times 0.1}{2 \times 9.81 \times (3 \times 10^{-6})^2 \times (2.65 - 1) \times 10^3} = 2752 \text{ s} = 45 \text{ m } 52 \text{ s}$$

3.6.3 Particle shape

Both these techniques – sieving and sedimentation – are describing the soil particles with reference to a simple model of equivalent spheres. It will probably be helpful, if the individual particles can be inspected with the naked eye or with a readily available optical microscope, to give some indication of the shape of the particles. Some indication of the scale of descriptors from *angular* to *well rounded* is hinted at in Fig. 3.16.

As a pedagogic laboratory exercise, it is good training for students to grasp the idea of combining objective quantitative measurements (for example, proportions retained on particular sieves) – which could be repeated by others with the expectation of achieving essentially the same numerical results – with descriptions (for example, particle shape or colour or general soil consistency) – which are more qualitative but nevertheless must be immediately clear to someone who is not actually present (*Think, when we talk of horses, that you see them printing their proud hoofs i'the receiving earth...*).

3.6.4 Sand: relative density

In Table 3.2 typical values for the densities of "dense" and "loose" soils were given. The terms "dense" and "loose" are rather vague terms which can be made slightly more objective if they are somehow tied to the range of densities for which a particular sandy or gravelly soil can exist. More or less standard laboratory tests[4] can be used to characterise this range – discovering a maximum density or minimum void ratio and a minimum density or maximum void ratio for the particular soil in its dry state.

A standard vibratory procedure is used to discover the minimum void ratio, e_{min}, of the soil. A series of repeated inversions of a large tube containing a sample of the sandy soil is used to estimate the maximum void ratio, e_{max}. The maximum and minimum void ratios determined in this way do not actually define the actual extremes of packing; they merely provide a useful index for the soil. In reality, the absolute lowest value of void ratio must be zero when the stresses have been increased so enormously that all the particles have broken and there are no longer any visible voids. And there is a maximum void ratio – a very loose packing – for which the particles are not really in proper stationary contact, and the material is not able to transmit stress from one side to the other.

However, given these index or reference values of void ratio, if the soil is prepared or found to exist at any other void ratio, e, then a relative density, D_r, can be defined:

$$D_r = \frac{e_{max} - e}{e_{max} - e_{min}} \tag{3.38}$$

This range of void ratios depends on the range of particle sizes and the typical particle shapes. For example, angular particles are able to exist in rather looser packings than rounded particles. Well-graded soils – soils such as the glacial till shown in Fig. 3.9 with a wide range of particle sizes – tend to have low void ratios because for any size of particles there exist smaller particles which are happy to sit in the spaces that would exist around the larger particles. The most efficient packing of particles would be linked with a self-similar fractal distribution of particle sizes in which the arrangement of particles has the same general appearance no matter at what scale the soil is observed. Then the proportion of particles within a range of sizes having the same ratio would be the same, so that if $d_1/d_2 = d_3/d_4$ then the proportion in the size range d_2 to d_1 would be the same as that in the size range d_4 to d_3, and so on.

Classification tests of this type are valuable in providing information which is expected to correlate, broadly, with stiffness and strength properties of the soil that may be relevant for calculation of the performance of geotechnical structures of which that soil forms part.

3.7 Summary

Here is a concise list of the key messages from this chapter, which are also encapsulated in the mind map (Fig. 3.17).

1. Soils are composed of mineral particles separated by voids which may be partially or wholly filled with liquid.

---

[4] Kolbuszewski, J.J. (1948) An experimental study of the maximum and minimum porosities of sands. *Proc. 2nd Int. Conf. on Soil mechanics and foundation engineering*, Rotterdam **1** 158–165.

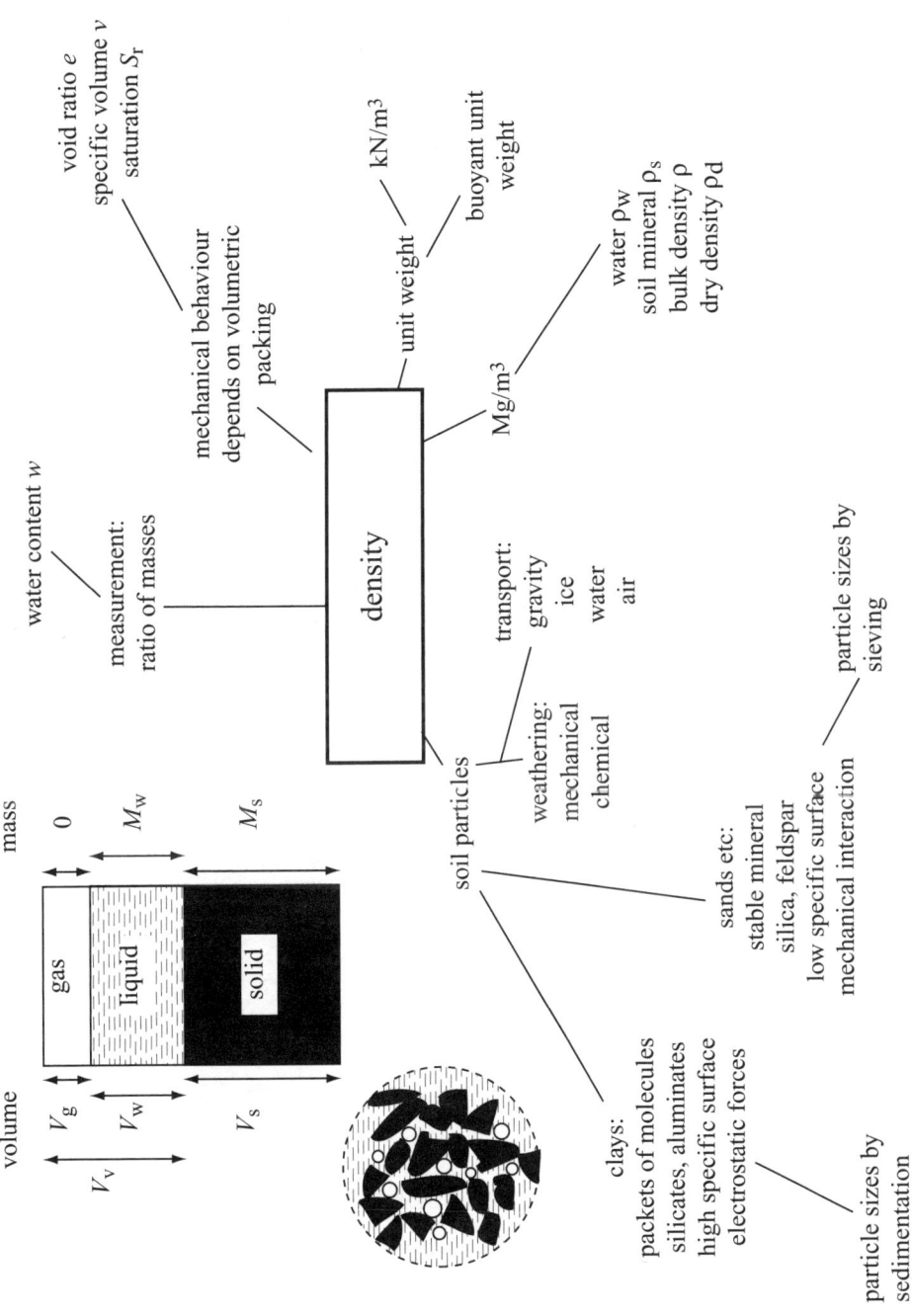

Figure 3.17. Mind map: density.

2. Many aspects of the behaviour of soils are strongly influenced by the density of packing of the mineral particles: it is necessary to define density or volumetric variables to characterise the packing.
3. Ratios of volumes are used to define void ratio, e, specific volume, v and porosity, n. The volumetric proportion of the voids filled with water is the degree of saturation, S_r.
4. Water content, w, is a ratio of masses and is straightforward to measure.
5. Determination of density requires some means of estimating the volume of a soil sample.
6. For a given amount of compaction energy, there is an optimum moisture content which produces the maximum density or minimum specific volume.
7. Soil particles originate from weathering of rocks and subsequent transport.
8. The sizes of particles that may be found in natural or man-made soils cover a very wide range.
9. The distribution of sizes of coarser particles can be found by sieving; the distribution of sizes of finer particles can be found by sedimentation.
10. Significant electrostatic forces between packets of clay molecules encourage open structures with high void ratios.
11. The larger inert grains of sands and gravels interact through interparticle contact forces.

3.8 Exercises: Density

3.8.1 Multiple choice questions

Choose the correct answer from the alternatives provided.

1. Void ratio $e =$ (a) $\frac{\text{volume of soil particles}}{\text{volume of voids}}$; (b) $\frac{\text{volume of voids}}{\text{total volume}}$; (c) $\frac{\text{volume of voids}}{\text{volume of soil particles}}$.
2. Specific volume $v =$ (a) $1+e$; (b) $1-e$; (c) $1/e$.
3. Water content $w =$ (a) $\frac{\text{volume of water}}{\text{volume of soil particles}}$; (b) $\frac{\text{mass of water}}{\text{total mass}}$; (c) $\frac{\text{mass of water}}{\text{mass of soil particles}}$.
4. Degree of saturation $S_r =$ (a) $\frac{\text{volume of water}}{\text{total volume}}$; (b) $\frac{\text{volume of water}}{\text{volume of voids}}$; (c) $\frac{\text{volume of water}}{\text{volume of soil particles}}$.
5. Density $\rho =$ (a) $\frac{\text{mass}}{\text{area}}$; (b) $\frac{\text{weight}}{\text{volume}}$; (c) $\frac{\text{mass}}{\text{volume}}$.
6. Dry density $\rho_d =$ (a) $G_s \rho_w$; (b) $\frac{G_s \rho_w}{1+e}$; (c) $\frac{\rho_w}{w}$.
7. Bulk unit weight $\gamma =$ (a) $(1+w)\rho_d$; (b) $(1+w)\rho_w$; (c) $(1+w)\rho_d g$.
8. Buoyant unit weight $\gamma' =$ (a) $\gamma - \gamma_w$; (b) $\gamma + \gamma_w$; (c) $\frac{\gamma}{\gamma_w}$.
9. $G_s = 2.7$, $w = 24\%$, $S_r = 100\%$.
   $e =$ (a) 0.548; (b) 0.648; (c) 0.748
   $v =$ (a) 1.548; (b) 1.748; (c) 1.648

3.8 Exercises: Density

$\rho_d =$ (a) 1.64 Mg/m³; (b) 0.64 Mg/m³; (c) 1.71 Mg/m³
$\rho_{sat} =$ (a) 2.25 Mg/m³; (b) 2.03 Mg/m³; (c) 1.98 Mg/m³
$\rho =$ (a) 1.9 Mg/m³; (b) 2.3 Mg/m³; (c) 2.03 Mg/m³
$\gamma =$ (a) 19.9 kN/m³; (b) 19.5 kN/m³; (c) 20.5 kN/m³
$\gamma_{sat} =$ (a) 19.0 kN/m³; (b) 19.9 kN/m³; (c) 20.5 kN/m³

10. $G_s = 2.64$, $w = 35\%$, $S_r = 75\%$
$e =$ (a) 1.232; (b) 1.105; (c) 1.302
$v =$ (a) 1.105; (b) 2.302; (c) 2.232
$\rho_d =$ (a) 1.18 Mg/m³; (b) 1.28 Mg/m³; (c) 1.81 Mg/m³
$\rho_{sat} =$ (a) 1.73 Mg/m³; (b) 1.63 Mg/m³; (c) 1.53 Mg/m³
$\rho =$ (a) 1.55 Mg/m³; (b) 1.6 Mg/m³; (c) 1.65 Mg/m³
$\gamma =$ (a) 15.7 kN/m³; (b) 15.1 kN/m³; (c) 16.3 kN/m³
$\gamma_{sat} =$ (a) 16.0 kN/m³; (b) 17.0 kN/m³; (c) 18.0 kN/m³

11. $M = 6$ kg, $V = 0.003$ m³, $M_s = 5$ kg, $G_s = 2.68$
$e =$ (a) 0.580; (b) 0.608; (c) 0.751
$v =$ (a) 1.608; (b) 1.580; (c) 1.751
$w =$ (a) 15%; (b) 25%; (c) 20%
$S_r =$ (a) 88%; (b) 83%; (c) 81%
$\rho_d =$ (a) 1.67 Mg/m³; (b) 1.75 Mg/m³; (c) 1.62 Mg/m³
$\rho_{sat} =$ (a) 2.15 Mg/m³; (b) 2.04 Mg/m³; (c) 2.26 Mg/m³
$\rho =$ (a) 1.9 Mg/m³; (b) 2 Mg/m³; (c) 2.2 Mg/m³
$\gamma =$ (a) 19.9 kN/m³; (b) 19.6 kN/m³; (c) 19.1 kN/m³
$\gamma_{sat} =$ (a) 20.2 kN/m³; (b) 20.0 kN/m³; (c) 20.5 kN/m³

3.8.2 Calculation exercises

1. A soil has bulk density $\rho = 1.8$ Mg/m³ and dry density $\rho_d = 1.5$ Mg/m³. The specific gravity of the soil particles is $G_s = 2.7$. Calculate the water content, specific volume, void ratio and degree of saturation.

2. A soil is found to have water content $w = 30\%$ and degree of saturation $S_r = 85\%$. The specific gravity of soil particles is $G_s = 2.65$. Calculate the void ratio, specific volume, dry density and bulk density.

3. A sample of soil of volume 0.0024 m³ is found to weigh 5.066 kg. After drying in an oven it is found to weigh 4.54 kg. The specific gravity of soil particles is $G_s = 2.7$. Calculate the bulk density, dry density, specific volume, void ratio, water content and degree of saturation.

4. A specimen of saturated soil (soil = soil particles + water) is placed in a measuring tin and weighed, then dried in an oven and then weighed again. The following data are recorded. Find the water content, void ratio and specific volume of the soil. Assume $G_s = 2.67$.

weight of tin	10.05 g
weight of tin + wet soil	35.74 g
weight of tin + dry soil	29.61 g

5. An undisturbed sample of saturated soil has a volume of 143 cm³ (143 × 10⁻⁶ m³) and a mass of 260 g. Determine its void ratio, specific volume, water content, density and dry density. Assume $G_s = 2.7$. (Hint: First find masses and volumes of soil and water components.)

6. A sample of soil has a bulk density of 1.9 Mg/m³ and a water content of 30%. Find the degree of saturation, void ratio, specific volume and dry density of the soil. Assume $G_s = 2.7$.

7. In a density determination, a sample of clay was weighed in air and its mass was found to be 683 g. It was coated with paraffin of specific gravity 0.89 to seal it. The combined mass of the clay and the wax was 690.6 g. To find its volume, the wax-coated specimen was suspended by a thread from a balance, immersed in water, and the balance read. Its apparent mass was found to have reduced to 340.6 g. The sample was then broken open and appropriate tests gave water content $w = 16.8\%$ and particle specific gravity $G_s = 2.73$. Determine the bulk density, void ratio, specific volume, degree of saturation and dry density of the clay. (Hint: Archimedes' principle indicates that the upthrust on a body is equal to the weight of water displaced by the body Section 2.7.)

8. Clay excavated from an underground tunnel of diameter 4.5 m under construction in central London is to be used to fill some old gravel workings near Heathrow. The gravel pits cover an area of 5000 m² and have an average depth of 10 m. Water fills the pits to just below the adjacent ground level. The clay will be transported by lorry and end tipped.

   The clay *in situ* is fully saturated, with water content 12% and specific gravity $G_s = 2.7$. Pilot tests show that, after tipping is completed, the uncompacted fill will remain saturated, but the average water content will rise to 23.5%. Estimate the length of tunnel needed to fill the pits and the number of 10 m³ lorry loads required. Assume the soil increases its volume by 15% when tipped into the lorry.

4 Stiffness

4.1 Introduction

The two principal mechanical properties of all materials that are required for engineering design are some way of knowing how strong the material is: how much stress it will tolerate – its strength – and some indication of the way in which it will change in size when subjected to load – its stiffness. These characteristics essentially form the basis of what are called, respectively, *ultimate limit state* design and *serviceability limit state* design. We will take a one-dimensional look at strength in Chapter 8. Here we will explore some aspects of stiffness of soils.

4.2 Linear elasticity

The standard experiment that can be performed on metal rods or wires to discover their deformation properties consists of the stretching of an appropriate specimen between suitable grips and measuring the link between the load applied and the resulting extension. In fact, the sort of experiment that can be performed at home might use a metal wire fixed to the ceiling and loaded by means of weights on a small pan (Fig. 4.1): the force transmitted to the wire is visibly obvious and, if we have sufficiently accurate position measuring devices – perhaps some optical system to magnify the displacement – we can have direct information about the extension as well (Fig. 4.1a).

For most metals, provided the loads that are applied are not excessive, the relationship between load, P, and extension, $\Delta\ell$, of such a wire is more or less linear (Fig. 4.1b). We can extract a material property, the elastic stiffness or *elastic modulus*, from the slope of this relationship. We could simply report the slope, $P/\Delta\ell$, but this would only provide a result relevant to the particular specimen of particular cross-section and particular length. We expect that the deformation resulting from the application of a given load will depend on the area of the cross-section over which it is spread. In Chapter 2 we introduced the concept of *stress* as an indication of the areal intensity of force. The axial stress in the wire of radius r will be $\sigma_a = P/\pi r^2$.

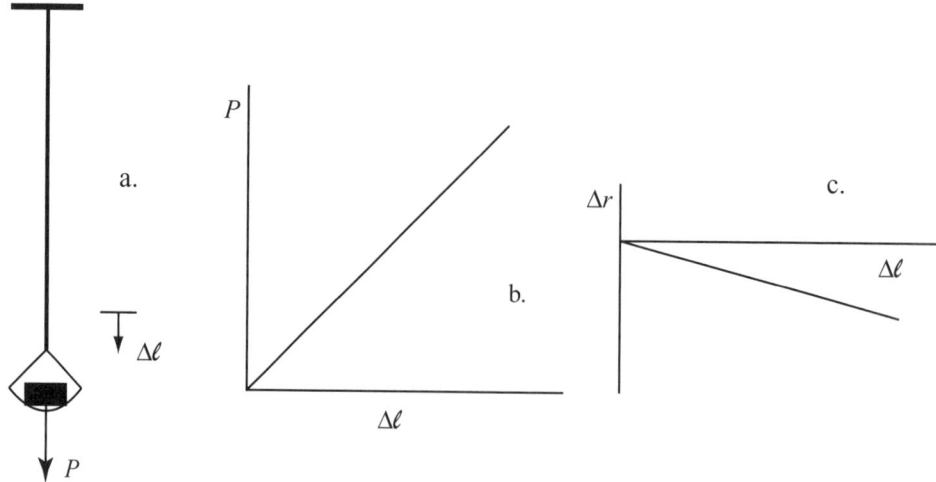

Figure 4.1. (a) Simple experiment to determine Young's modulus of metal wire; (b) linear elastic relationship between load P and extension $\Delta\ell$; (c) radial contraction Δr accompanying axial extension $\Delta\ell$.

But we need also to normalise the extension. Again our intuition suggests that, if we apply a certain load to a wire of a certain cross section, then the longer the wire the greater will be the extension. We define axial *strain* in the wire, ε_a, as the ratio between the extension $\Delta\ell$ and the length ℓ of the wire: $\varepsilon_a = \Delta\ell/\ell$. Then the ratio between the axial stress and the axial strain gives us a much more useful and transferable material property which we may suppose can be applied to wires (or other forms of sample of the material) of other lengths and other cross sections. This property is *Young's modulus E*:

$$E = \frac{P}{\pi r^2} \frac{\ell}{\Delta\ell} = \frac{\sigma_a}{\varepsilon_a} \qquad (4.1)$$

Because axial strain is a dimensionless quantity, Young's modulus has the dimensions of stress and is reported in kilopascals (1 kPa = 10^3 Pa), megapascals (1 MPa = 10^6 Pa) or gigapascals (1 GPa = 10^9 Pa) as appropriate. Thus Young's modulus for steel is of the order of 210 GPa.

In his compendious but challenging *A course of lectures on natural philosophy and the mechanical arts* published in 1807, Thomas Young introduces Young's modulus in these terms: "we may express the elasticity of any substance by the weight of a certain column of the same substance, which may be denominated the modulus of its elasticity, and of which the weight is such, that any addition to it would increase it in the same proportion as the weight added would shorten, by its pressure, a portion of the substance of equal diameter". Imagine a column of material of height H and cross sectional area A. We add a little more material to the column, increasing its height by δH, but the increase in weight of the column causes an elastic shortening exactly equal to δH and the height remains constant. The increase in weight is $\rho g A \delta H$, the increase in stress is $\rho g \delta H$, the axial strain is $\rho g \delta H/E$ and

4.2 Linear elasticity

the shortening of the column is $\rho g H \delta H/E$. Hence the height of the column is $H = E/\rho g$ and we can quote values of Young's modulus (as Thomas Young does) in units of length! Young quotes the modulus of steel as about 1500 miles (2400 km). The density of steel is about 7.8 Mg/m^3 so 2400 km corresponds to 184 GPa – a bit low (Young's modulus for steel is usually around 200 GPa) but of the correct order of magnitude. A closer estimate would be 2600 km (or 1600 miles).

The observation of the extension of the wire is more or less straightforward. However, it is again not particularly surprising that, as we pull the wire and make it longer, it also becomes narrower and, with a sufficiently sensitive micrometer, we could measure the change in radius Δr, which we would discover also varied linearly with the applied load (Fig. 4.1c). Converting this change in radius to a radial strain, $\varepsilon_r = \Delta r/r$, we could determine a second elastic property of the metal. The ratio of radial to axial strain is called *Poisson's ratio* ν:

$$\nu = -\frac{\Delta r/r}{\Delta \ell/\ell} = -\frac{\varepsilon_r}{\varepsilon_a} \tag{4.2}$$

the negative sign being needed because the radial strain will be compressive if the axial strain is tensile and *vice versa*. We could also see Poisson's ratio as an indication of an unexpected coupled stiffness: we have pulled the wire in the axial direction but we are observing deformations in the radial direction even though we are applying no stresses in this direction. We could express this unexpected stiffness as a proportion of Young's modulus:

$$\frac{P}{\pi r^2} \frac{r}{\Delta r} = \frac{\sigma_a}{\varepsilon_r} = -\frac{E}{\nu} \tag{4.3}$$

The value of Poisson's ratio tells us something about the change in volume of the specimen that occurs as it is being pulled. The volume of the wire is $V = \pi r^2 \ell$. If the radius changes to $r + \Delta r$, where $\Delta r \ll r$, and the length changes to $\ell + \Delta \ell$ where $\Delta \ell \ll \ell$, the new volume of the wire will be $V + \Delta V$:

$$\begin{aligned} V + \Delta V &= \pi(r + \Delta r)^2(\ell + \Delta \ell) = \pi r^2 \ell + \pi r^2 \Delta \ell + 2\pi r \ell \Delta r \\ &= V + \pi r^2 \Delta \ell + 2\pi r \ell \Delta r \end{aligned} \tag{4.4}$$

neglecting second order small quantities (terms involving products of Δr and $\Delta \ell$). Calculating a volumetric strain ε_{vol} in the same way that we calculated axial and radial strain – as the ratio of change in volume to original volume – and introducing the definitions of axial and radial strain and Poisson's ratio, we find:

$$\varepsilon_{vol} = \frac{\Delta V}{V} = \frac{\Delta \ell}{\ell} + 2\frac{\Delta r}{r} = (1 - 2\nu)\varepsilon_a \tag{4.5}$$

We can deduce the general result that a material which has Poisson's ratio $\nu = 0.5$ deforms at constant volume.

4.3 Natural and true strain

We have not placed any restrictions on the magnitudes of the stresses or strains. In general, we will be interested to estimate the response of our soil to a change in stress: what will then be the resulting change in strain? There are some subtleties concerning the definition of strain. What is strain? We have defined axial strain as proportional change in length, but we have to choose a length against which to compare the change. There are two ways in which we can do this. A small (infinitesimal) increment of natural strain $\delta\varepsilon_{zn}$ compares the change in length $\delta\ell$ with the initial value ℓ_o. So for vertical strain, in the z direction:

$$\delta\varepsilon_{zn} = \frac{\delta\ell}{\ell_o} \qquad (4.6)$$

and

$$\varepsilon_{zn} = \int_0^{\varepsilon_{zn}} d\varepsilon_{zn} = \int_{\ell_o}^{\ell} \frac{d\ell}{\ell_o} = \frac{\ell - \ell_o}{\ell_o} = \frac{\Delta\ell}{\ell_o} \qquad (4.7)$$

where $\Delta\ell$ is the eventual, non-infinitesimal, change in length.

On the other hand, the increment of true strain $\delta\varepsilon_{zt}$ compares the change in length with the current length ℓ so that:

$$\delta\varepsilon_{zt} = \frac{\delta\ell}{\ell} \qquad (4.8)$$

and

$$\varepsilon_{zt} = \int_0^{\varepsilon_{zt}} d\varepsilon_{zt} = \int_{\ell_o}^{\ell} \frac{d\ell}{\ell} = \ln\frac{\ell}{\ell_o} = \ln\frac{\ell_o + \Delta\ell}{\ell_o} = \ln\left(1 + \frac{\Delta\ell}{\ell_o}\right) = \ln(1 + \varepsilon_{zn}) \qquad (4.9)$$

The infinite series expansion of $\ln(1 + y)$ is:

$$\ln(1 + y) = y - \frac{y^2}{2} + \frac{y^3}{3} - \frac{y^4}{4}\ldots \qquad (4.10)$$

If the changes in length are small, $\Delta\ell \ll \ell_o$, then, comparing (4.9) and (4.10) and (4.7):

$$\varepsilon_{zt} \approx \frac{\Delta\ell}{\ell_o} = \varepsilon_{zn} \qquad (4.11)$$

and the difference between these two definitions of strain is also small. But, as shown in Fig. 4.2, as the change in length becomes greater the difference also increases. In compression, true strain is greater than natural strain; in extension, true strain is smaller than natural strain.

4.4 One-dimensional testing of soils

Tensile testing of soil specimens in the form of a wire or other shape is not usually practicable – imagine trying to pull a sample of sand – though it is possible to machine appropriate geometries from some rocks. We will expect instead to be performing *compression* tests so that the sample becomes shorter and fatter as the load

4.4 One-dimensional testing of soils

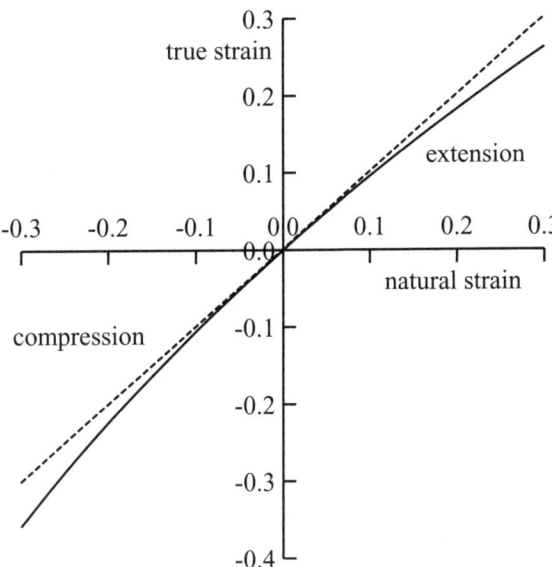

Figure 4.2. Comparison of natural and true strains for uniaxial compression and extension (dotted line indicates identity).

is increased (Fig. 4.3). Exactly the same sort of interpretation in terms of Young's modulus and Poisson's ratio can be applied – the signs of both the load and the dimensional changes or strains will be reversed.

Restricting ourselves to a one-dimensional form of loading, we will have to limit ourselves (for the moment) to a special but common form of compression testing in which no lateral strain of the soil is permitted: compare the loading arrangements shown schematically in Fig. 4.4. The actual device that is used in the soil mechanics laboratory for this testing is called an *oedometer* (Fig. 4.5). The sample of soil is contained in a stiff (rigid?) ring which prevents any change in cross-section.

Now the effect of confining the soil will be to increase the stiffness for our one-dimensional loading configuration. We know that the Poisson's ratio effect will make the soil want to expand sideways as it is compressed vertically but the

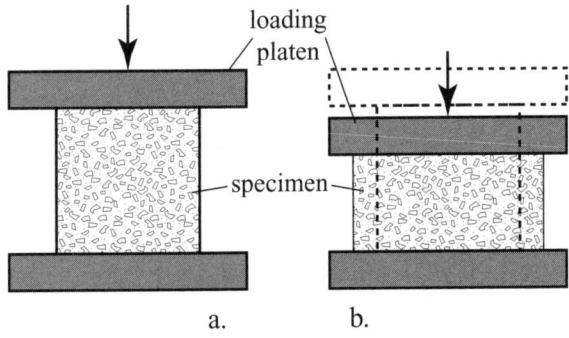

Figure 4.3. Compression test on unconfined specimen.

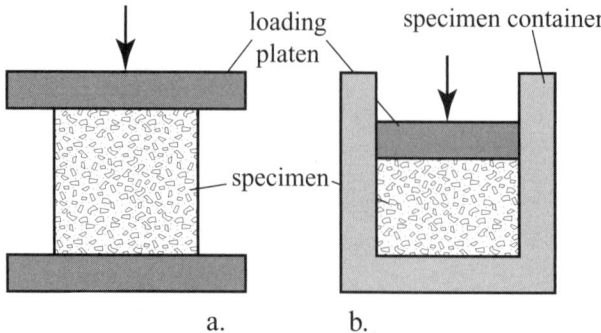

Figure 4.4. (a) Compression test on unconfined specimen; (b) compression test on confined specimen.

containment ring will prevent this (think about the constraint provided in Fig. 4.4b by comparison with the freedom evident in Fig. 4.3). As the load is increased, lateral stresses will develop and will tend, again through the restrained Poisson's ratio effect, to push the load upwards or to reduce the vertical strain. The stiffness for one-dimensional elastic *confined* compression, E_o, will be greater than the Young's modulus, E, which is the stiffness for one-dimensional *unconfined* compression. We can express this confined stiffness, E_o, in terms of Young's modulus and Poisson's ratio (Fig. 4.6):

$$E_o = \frac{\sigma_z}{\varepsilon_z} = E\frac{(1-\nu)}{(1-2\nu)(1+\nu)} \tag{4.12}$$

but the proof of this expression requires us to escape temporarily from our one-dimensional constraint and is contained in a subsection, which can be omitted if desired.

4.4.1 Hooke's Law: confined one-dimensional stiffness ♣

The name of Robert Hooke is associated with the proposal of the linear elastic relationship between stress and strain, originally published in the form of a Latin anagram which could be interpreted as *ut tensio sic vis* – "as the extension, so the

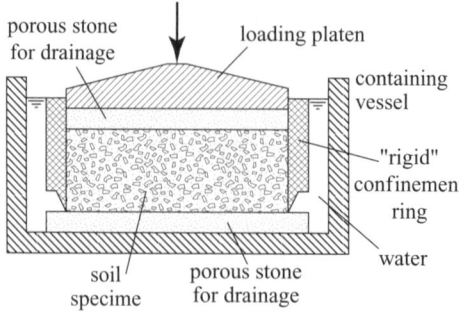

Figure 4.5. Oedometer: confined one-dimensional compression test.

4.4 One-dimensional testing of soils

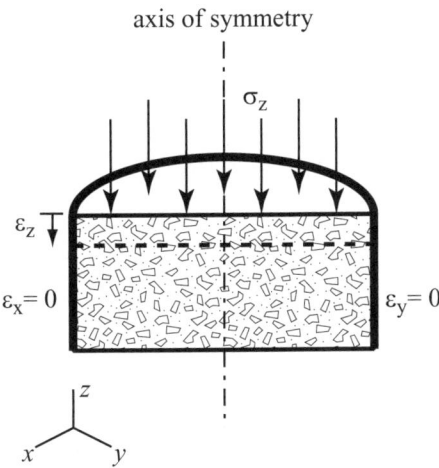

Figure 4.6. One-dimensional compression of soil specimen.

force" – which is exactly the relationship sketched in Fig. 4.1. Analysis of stresses in any continuous medium indicates that any general stress state can always be reduced to a set of three mutually orthogonal *principal* stresses (Fig. 4.7). For a linear elastic material, we can superpose results for different stress states and can build up the elastic response for this three-dimensional principal stress state from what we already know about our one-dimensional wire.

Consider the effect of applying, on its own, the stress σ_x in the x direction. Young's modulus tells us about the strains in the x direction; Poisson's ratio tells us about the strains in the orthogonal y and z directions. The resulting strains are:

$$\varepsilon_x = \frac{\sigma_x}{E}; \qquad \varepsilon_y = -\nu\frac{\sigma_x}{E}; \qquad \varepsilon_z = -\nu\frac{\sigma_x}{E} \qquad (4.13)$$

Now consider the effect of applying, on its own, the stress σ_y. The same principle applies and the resulting strains are:

$$\varepsilon_x = -\nu\frac{\sigma_y}{E}; \qquad \varepsilon_y = \frac{\sigma_y}{E}; \qquad \varepsilon_z = -\nu\frac{\sigma_y}{E} \qquad (4.14)$$

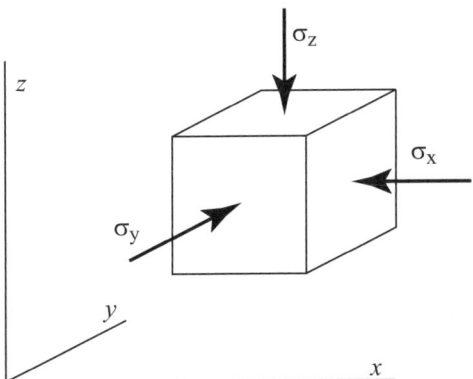

Figure 4.7. General stress state described by three mutually orthogonal principal stresses.

Finally, consider the effect of applying, on its own, the stress σ_z. The resulting strains are:

$$\varepsilon_x = -\nu\frac{\sigma_z}{E}; \qquad \varepsilon_y = -\nu\frac{\sigma_z}{E}; \qquad \varepsilon_z = \frac{\sigma_z}{E} \qquad (4.15)$$

Then we superpose all these strains to find the effect of applying all three principal stresses:

$$\varepsilon_x = \frac{\sigma_x}{E} - \nu\frac{\sigma_y}{E} - \nu\frac{\sigma_z}{E}; \quad \varepsilon_y = -\nu\frac{\sigma_x}{E} + \frac{\sigma_y}{E} - \nu\frac{\sigma_z}{E}; \quad \varepsilon_z = -\nu\frac{\sigma_x}{E} - \nu\frac{\sigma_y}{E} + \frac{\sigma_z}{E} \qquad (4.16)$$

and this is Hooke's Law of linear elasticity stated in terms of principal stresses and strains.

Now the one-dimensional constrained compression of the elastic material in the oedometer is just a special case of this general elastic stress condition. We impose two conditions:

- The stress state has a symmetry about the vertical z axis (Fig. 4.6) so that the stresses in the x and y directions must be identical $\sigma_x = \sigma_y$.
- The constrained nature of the compression requires that the strains in the x and y directions must both be zero, $\varepsilon_x = \varepsilon_y = 0$.

Applying these two conditions to (4.16) we find first that:

$$\sigma_x = \sigma_y = \frac{\nu}{1-\nu}\sigma_z \qquad (4.17)$$

and second, after a little manipulation, having substituted these values of σ_x and σ_y, that the one-dimensional confined stiffness, E_o, is:

$$E_o = \frac{\sigma_z}{\varepsilon_z} = E\frac{(1-\nu)}{(1-2\nu)(1+\nu)} \qquad (4.18)$$

and it should not surprise us that, if Poisson's ratio $\nu = 0.5$, which we saw implied constant volume elastic response, then the one-dimensional stiffness E_o is infinite – the material cannot expand laterally and therefore cannot compress axially either. The more general conclusion that we can draw is that, if we observe stiffness in this way, we will only be able to discover the composite elastic property, E_o (4.18), and will not be able to discover the separate values of Young's modulus and Poisson's ratio. The relationship between the ratio E_o/E and Poisson's ratio ν is shown in Fig. 4.8.

4.5 One-dimensional (confined) stiffness of soils

Some examples of one-dimensional compression of soils are shown in Fig. 4.9, plotted in terms of vertical stress and specific volume rather than volumetric or vertical strain. Two characteristics of these responses are immediately evident: the compression is very nonlinear and it is irreversible. We will discuss these two features separately but there is one immediate consequence of the combination of the two characteristics: we have to recognise that the stiffness that we observe as an incremental response is not an elastic stiffness. For a truly elastic material, the behaviour

4.5 One-dimensional (confined) stiffness of soils

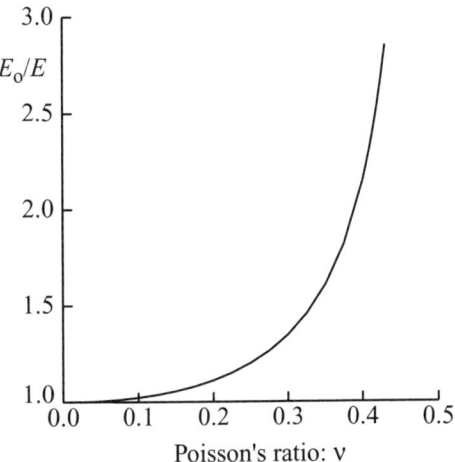

Figure 4.8. Ratio of confined modulus and Young's modulus as function of Poisson's ratio ν.

on loading and unloading would be the same and we would recover the same density of packing when we found ourselves applying once again the same load. So a word of warning is needed: geotechnical engineers are a bit slack in using the language of elasticity to describe behaviour which is certainly not elastic in the strict sense. A material that is linear and elastic follows the principle of superposition (this was implicit in the way in which we made use of Hooke's Law, Section 4.4.1), and this will not be strictly possible for real soils.

A thought experiment concerning the one-dimensional compression of soil is shown in Fig. 4.10. This is a completely generic picture which indicates our expectation that stiffness will increase as the density of the soil increases or its specific

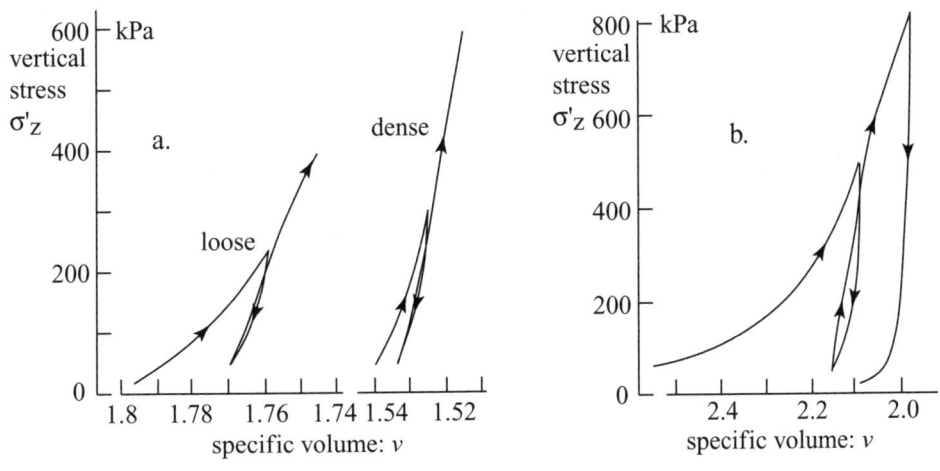

Figure 4.9. One-dimensional (confined) compression of (a) sand and (b) clay.

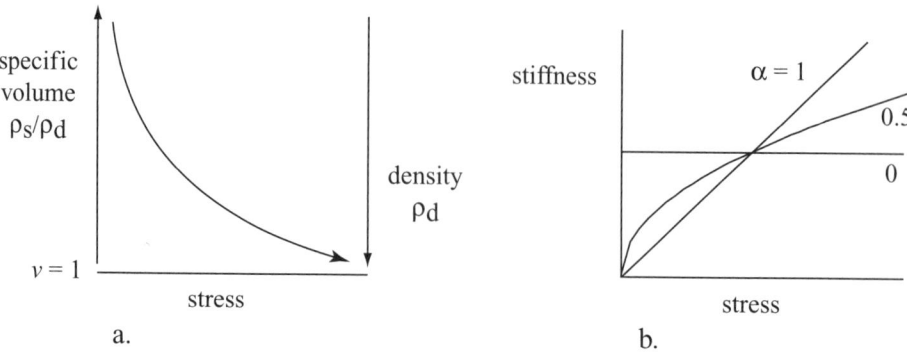

Figure 4.10. One-dimensional (confined) compression of soil.

volume or void ratio reduces. Janbu (1963)[1] suggested that a power law might be used to describe the link between incremental stiffness E_o and vertical effective stress σ'_z:

$$\frac{E_o}{\sigma_{ref}} = \chi \left(\frac{\sigma'_z}{\sigma_{ref}}\right)^\alpha \qquad (4.19)$$

Recall that when we introduced the concept of effective stress in Section 2.8 we noted that effective stresses were important because they controlled all aspects of stiffness and strength of soils. Here is an example of their role in controlling stiffness.

The reference stress σ_{ref} is introduced to leave dimensional consistency between the two sides of (4.19). Atmospheric pressure is often used as just such a reference stress. It has the value of about 100 kPa and its introduction at least has the benefit of ensuring that the units used for stiffness and for stresses are compatible. On the other hand, it is logical to remark that the reference stress should really be some material property which has some influence on the observed response. McDowell[2] suggests that much of the nonlinearity observed in compression of sands is associated with breakage of particles; that the breakage of particles is largely controlled by the tensile strength of the mineral of which the particles are composed; and that the mineral tensile strength provides a more appropriate reference stress to use in (4.19). That is a subject of continuing research – we here take the value of $\sigma_{ref} = 100$ kPa.

The value of the exponent α in (4.19) controls the way in which the incremental stiffness changes with vertical stress σ'_z. If $\alpha = 0$, then the stiffness is constant (Fig. 4.10b), $E_o = \chi \sigma_{ref}$, and does not change with stress level. This returns us to linear elasticity. Such a value of α might be appropriate for rocks.

If $\alpha = 1$, then the stiffness is directly proportional to the vertical stress (Fig. 4.10b). Some indication of the range of values for different soil types is shown in

---

[1] Janbu, N. (1963) Soil compressibility as determined by oedometer and triaxial tests. *Proceedings of 3rd European Conference on Soils Mechanics and Foundation Engineering*, Wiesbaden, Germany **1** 19–25.

[2] McDowell, G.R., Bolton, M.D. & Robertson, D. (1996) The fractal crushing of granular materials. *J Mech. Phys. Solids* **44** 12, 2079–2102.

4.5 One-dimensional (confined) stiffness of soils

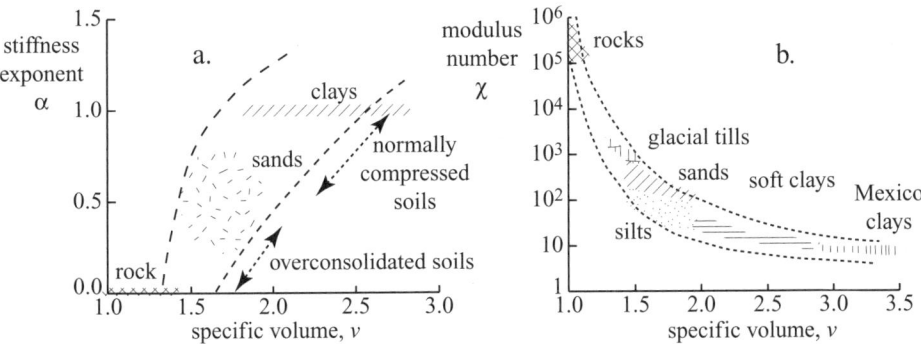

Figure 4.11. Stiffness parameters for geomaterials: (a) α and (b) χ (adapted from Janbu, 1963).[1]

Fig. 4.11a. There is a general broad correlation with specific volume v which, it will be recalled, is the inverse of volume fraction: the greater the proportion of volume occupied by solid mineral (the smaller the value of v), the lower the value of α and the nearer the one-dimensional stiffness approaches the linear elastic behaviour. When the soil has been so heavily compressed that there are no voids left, then $v = 1$ and we should recover the rock-like properties of the pure soil mineral.

The second material parameter in (4.19) is the modulus number χ, which indicates the magnitude of the stiffness at the reference stress level. Some general indication of the range of values of χ is shown in Fig. 4.11b revealing again, as one might expect, a general broad correlation between this reference stiffness and the specific volume or volume fraction. The values of χ are indicated on a logarithmic scale: the range covers several orders of magnitude.

The nonlinearity of stiffness apparent in the changing gradients of the one dimensional compression curves in Figs 4.9 and 4.10 has led people to seek ways of linearising the behaviour. The standard replotting uses a logarithmic scale for the stress axis (Fig. 4.12b). The logarithmic function stretches low values and

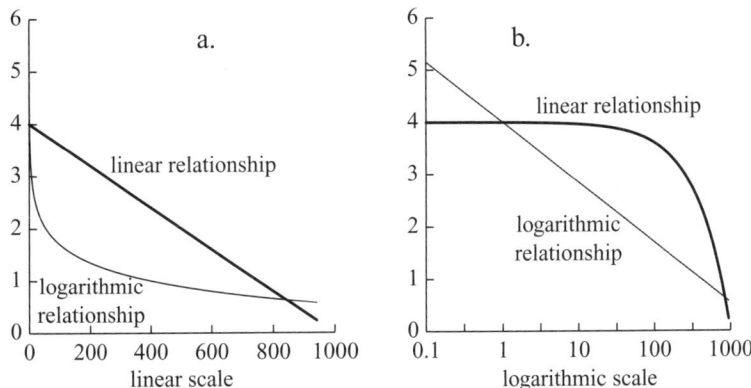

Figure 4.12. Illustration of effect of semi-logarithmic plotting of data. Linear ($y = 4 - 0.004x$) and logarithmic ($y = 4 - 0.5 \ln x$) relationships plotted with (a) linear x axis and (b) logarithmic x axis.

compresses high values and is often an effective way of removing curvature from observed patterns of response.

If we choose a value of $\alpha = 1$ in (4.19), then the one-dimensional stiffness becomes:

$$E_o = \frac{\delta \sigma_z'}{\delta \varepsilon_z} = \chi \sigma_z' \qquad (4.20)$$

so that the increment of axial strain:

$$\delta \varepsilon_z = \frac{\delta \sigma_z'}{E_o} = \frac{\delta \sigma_z'}{\chi \sigma_z'} \qquad (4.21)$$

We can integrate this to give:

$$\varepsilon_z = \int \frac{d\sigma_z'}{\chi \sigma_z'} = \frac{1}{\chi} \ln \frac{\sigma_z'}{\sigma_{zo}'} \qquad (4.22)$$

where σ_{zo}' is an initial value of vertical stress at which the strain is deemed to be zero. Thus this value of $\alpha = 1$ is directly compatible with the logarithmic plotting of stress which, from Fig. 4.11, we can expect to be particularly appropriate for presenting the behaviour of clays. However, the actual range of values of α should remind us that this logarithmic plotting of stress may not always be helpful. It is always tempting to try to fit the same simple model to materials who do not think that they know of its existence.

When we work in terms of specific volumes, volumetric strain in the oedometer is the same as vertical strain because there is no lateral strain which is able to contribute to the change in volume: $\delta \varepsilon_z = \delta \varepsilon_{vol} = -\delta v / v$ (keeping a sign convention of positive strains in compression). If we choose to work with the true strain, then we recall the logarithmic definition of strain from (4.9) and (4.7):

$$\varepsilon_{zt} = -\ln \frac{\ell}{\ell_o} = -\ln \frac{v}{v_o} = \ln v_o - \ln v \qquad (4.23)$$

We might choose to plot the observed response in terms of true strain and logarithm of stress to straighten out the experimental response. This would then be an application of a double logarithmic treatment of the data of specific volume and vertical stress.

4.6 Calculation of strains

We have an expression for the variation of stiffness with stress (4.19) and we can integrate this equation to discover the strain in an element of soil, one-dimensionally confined, caused by an increase in stress, and hence we can calculate the resulting change in dimension of the element of soil. The increment of strain, $\delta \varepsilon_z$, is the ratio of the increment of effective stress, $\delta \sigma_z'$, and the stiffness, E_o:

$$\delta \varepsilon_z = \frac{\delta \sigma_z'}{E_o} \qquad (4.24)$$

4.6 Calculation of strains

In the limit we can use the integral to sum the effect of increasing the stress in infinitesimal steps from the initial stress σ'_{z1} to the final stress σ'_{z2}:

$$\varepsilon_z = \int_0^{\varepsilon_z} \mathrm{d}\varepsilon_z = \int_{\sigma'_{z1}}^{\sigma'_{z2}} \frac{\mathrm{d}\sigma'_z}{E_o}$$

$$= \frac{1}{\chi \sigma_{ref}^{1-\alpha}} \int_{\sigma'_{z1}}^{\sigma'_{z2}} \frac{\mathrm{d}\sigma'_z}{\sigma'^{\alpha}_z} \qquad (4.25)$$

$$= \frac{1}{\chi \sigma_{ref}^{1-\alpha}} \left[\frac{1}{1-\alpha} \sigma'^{1-\alpha}_z \right]_{\sigma'_{z1}}^{\sigma'_{z2}}$$

$$\varepsilon_z = \frac{1}{\chi(1-\alpha)} \left[\left(\frac{\sigma'_{z2}}{\sigma_{ref}} \right)^{1-\alpha} - \left(\frac{\sigma'_{z1}}{\sigma_{ref}} \right)^{1-\alpha} \right]$$

This solution breaks down for the particular case when $\alpha = 1$, which produces the logarithmic relationship described in the previous section. For $\alpha = 1$:

$$\varepsilon_z = \frac{1}{\chi} \int_{\sigma'_{z1}}^{\sigma'_{z2}} \frac{\mathrm{d}\sigma'_z}{\sigma'_z}$$

$$= \frac{1}{\chi} \ln \frac{\sigma'_{z2}}{\sigma'_{z1}} \qquad (4.26)$$

Then, knowing the strain, we can calculate the change in thickness, or settlement ς of a soil layer of original thickness ℓ_o. If the strain is small, then $\varepsilon_z = (\ell_o - \ell)/\ell_o = \varsigma/\ell_o$ and $\varsigma = \ell_o \varepsilon_z$ and we remind ourselves that with our chosen sign convention of positive strains in compression, the soil layer is becoming thinner so that $\ell < \ell_o$. If the strain is large, then we may feel more comfortable using the true strain so that $\varepsilon_z = -\ln \ell/\ell_o$ and $\varsigma = \ell_o[1 - \exp(-\varepsilon_z)]$.

4.6.1 Worked examples: Calculation of settlement

1. As an example, consider the site conditions shown in Fig. 4.13. The ground consists of several layers of different soil types but there is, in particular, a soft layer some 2 m thick. The site is being prepared for a major construction project which requires the placement of fill on the surface, which will increase the stresses by 45 kPa. There is concern that the settlements that will develop in the soft layer may be substantial.

   In order that we may calculate the settlement, we need to know the stiffness properties of the soft layer. Laboratory testing has shown that it is characterised by a stiffness exponent $\alpha = 1$ and a modulus number $\chi = 10$. We also need to know the initial (effective) stresses in the soft layer. For simplicity, we will take an average value of initial vertical stress of $\sigma_{z1} = 60$ kPa (though one could clearly perform a slightly more complicated calculation in which the variation of stress within the layer was taken into account). Then the final stress in the layer will be $\sigma_{z2} = 60 + 45 = 105$ kPa and we can compute the vertical strain

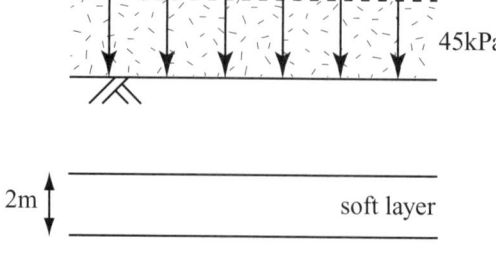

Figure 4.13. Site conditions for calculation of settlement.

from (4.26):
$$\varepsilon_z = \frac{1}{10} \ln \frac{105}{60} = 0.056 \quad (4.27)$$

Thus the placement of the fill produces a vertical strain in the soft layer of some $\varepsilon_z = 5.6\%$. The layer has a thickness of $\Delta z = 2$ m so that the vertical settlement in that layer is $\varsigma = \varepsilon_z \Delta z = 112$ mm.

2. More generally, if we are interested in summing the settlements from a number n of layers of thickness Δz_i (Fig. 4.14 – and one way of dealing with the variation of initial stress within a single soil layer is to divide it into thinner sub-layers which can individually be treated as uniform), then this calculation is repeated for each layer and the total settlement ς is deduced by summation.

$$\varsigma = \sum_{i=1}^{n} \varepsilon_{zi} \Delta z_i \quad (4.28)$$

Figure 4.15 shows a site where 6 m of soft soil overlie rock. The soft soil has unit weight $\gamma = 17$ kN/m³ and is characterised by values of stiffness exponent $\alpha = 0.8$ and modulus number $\chi = 15$. The water table is at the ground surface. The site is being prepared by placing a layer of fill generating a uniform increase in stress of $\Delta \sigma_z = 25$ kPa. We want to calculate the resulting settlement. We will

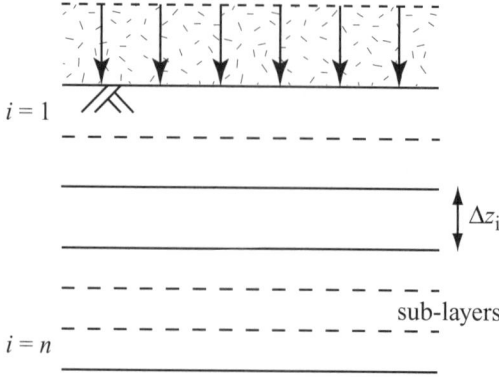

Figure 4.14. Layered site conditions for calculation of settlement.

4.6 Calculation of strains

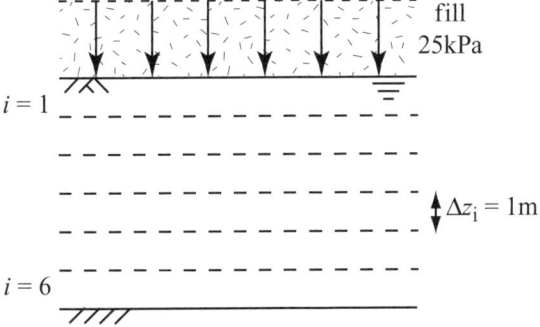

Figure 4.15. Placement of fill on soft soil: division of soil into layers for calculation of settlement.

divide the soft soil into six layers, each of thickness 1 m, to take account of the varying initial stress state in the ground.

For each layer, we need to calculate the initial effective stress σ'_{z1} from the given information about soil unit weight and water table. For example, for the layer between 2 m and 3 m, the total stress is $\sigma_z = 2.5 \times 17 = 42.5$ kPa; the pore pressure is $u = 2.5 \times 9.81 = 24.525$ kPa; and the initial effective stress is $\sigma'_{z1} = \sigma_z - u = 42.5 - 24.525 = 17.975$ kPa. Then we can calculate the final effective stress $\sigma'_{z2} = \sigma'_{z1} + \Delta\sigma_z$. For our layer between 2 m and 3 m, $\sigma'_{z2} = 17.975 + 25 = 42.975$ kPa. We can make use of (4.25) to calculate the strain in each layer and hence the change in thickness and the contribution to the settlement. For the layer between 2 m and 3 m, the strain is estimated to be 4.5% so that the settlement of that layer of thickness 1 m is 0.045 m. Finally, we sum the settlements from each layer to find the total settlement resulting from the placement of the fill. The results of the calculations are summarised in Table 4.1 for each of the six layers: the overall settlement is estimated to be 0.289 m, and the individual contributions from the 1 m layers vary from 0.088 m in the top layer to 0.028 m in the lowest layer simply because of the way in which the stiffness increases with stress level.

Table 4.1. *Calculation of settlement resulting from placement of fill as shown in Fig. 4.15.*

Layer	Depth m	σ_z kPa	u kPa	σ'_{z1} kPa	σ'_{z2} kPa	ε_z %	ς m
1	0–1	8.5	4.905	3.595	28.595	8.8	0.088
2	1–2	25.5	14.715	10.785	35.785	5.8	0.057
3	2–3	42.5	24.525	17.975	42.975	4.5	0.045
4	3–4	59.5	34.335	25.165	50.165	3.7	0.037
5	4–5	76.5	44.145	32.355	57.355	3.2	0.032
6	5–6	93.5	53.955	39.545	64.545	2.8	0.028
Total							**0.289 m**

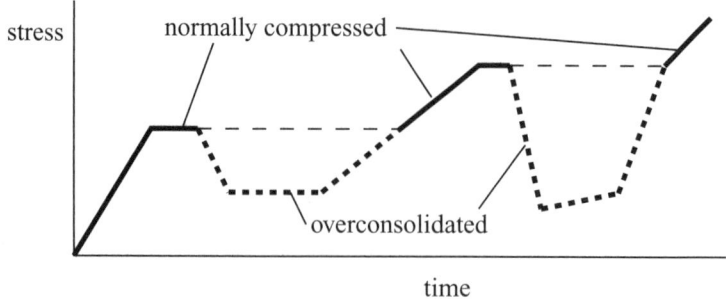

Figure 4.16. History of stress changes: normally compressed (current stress = past maximum stress: solid line) and overconsolidated (current stress < past maximum stress: dotted line).

4.7 Overconsolidation

In presenting some examples of one-dimensional compression of soils in Fig. 4.9 we noted that the compression was very nonlinear and that it was irreversible. We have discussed the nonlinearity; now we need to deal with the irreversibility.

A *normally compressed* (or normally consolidated) soil is one that has never experienced a vertical stress greater than the present one (Fig. 4.16). So long as the load is continually increasing, the increase of one-dimensional stiffness might be described by the relationship (4.19) as the density of the soil increases. However, when the rate of loading is reversed and the vertical stress starts to be removed much of the density change that has occurred is locked into the soil and the stiffness is much higher at any particular stress: the density change is thus irreversible (Fig. 4.17). This irreversibility of deformation is another indication that the framework of elasticity (Hooke's Law, Young's modulus, Poisson's ratio) is going to be inadequate to describe the deformations of soils. Even if it is nonlinear, elasticity does expect the deformations to be recovered when the load is reversed.

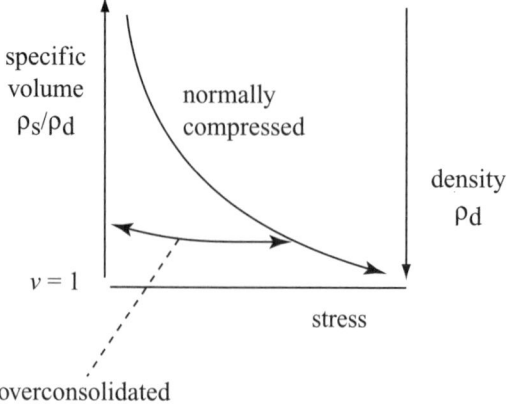

Figure 4.17. One-dimensional compression of normally compressed and overconsolidated soil.

4.7 Overconsolidation

When the soil is subjected to a vertical effective stress lower than the maximum effective stress that the soil has experienced in the past, it is said to be *overconsolidated*, and an overconsolidation ratio (n or ocr) can be defined:

$$n = \text{ocr} = \frac{\sigma'_{z_{max}}}{\sigma'_z} \qquad (4.29)$$

So far as the stiffness of overconsolidated soils is concerned, we may be able more closely to describe it as (nonlinear) elastic during episodes of unloading and reloading below the past maximum stress. It is true that the experimental observations in Fig. 4.9 show that the unloading and reloading tracks are not quite the same but the difference may be considered small in comparison with the much larger deformations that have accompanied the preceding normal compression. The stiffness will probably still vary with stress – the unloading and reloading paths in Fig. 4.9 are not straight – but the modulus number χ in (4.19) will depend on the density of the soil and the exponent α may be somewhat lower – particularly when the degree of overconsolidation is high. These characteristics are hinted at in Fig. 4.11.

We can encapsulate our stiffness model in three criteria:

1. If $\sigma'_z = \sigma'_{z_{max}}$ and $\delta\sigma'_z > 0$, then $n = 1$ and $\alpha = \alpha_{nc}$ and $\chi = \chi_{nc}$.
2. If $\delta\sigma'_z < 0$, then $n \geq 1$, $\delta n > 0$ and $\alpha = \alpha_{oc}$ and $\chi = \chi_{oc}$.
3. If $\sigma'_z < \sigma'_{z_{max}}$ and $\delta\sigma'_z > 0$, then $n > 1$ and $\alpha = \alpha_{oc}$ and $\chi = \chi_{oc}$.

The first criterion tells us that when the soil is normally compressed, so that $n = 1$, then the modulus number χ and exponent α take their normally compressed values, χ_{nc} and α_{nc}, provided the stress change is positive so that the maximum stress is always equal to the current stress.

The second criterion tells us that if the stress is being reduced, then we must either already be or else be becoming overconsolidated because the maximum stress $\sigma'_{z_{max}}$ is not changing. The modulus number χ and exponent α take their overconsolidated values, χ_{oc} and α_{oc}.

The third criterion tells us that, even if the stress increment is positive, then, provided the stress remains below the past maximum value, the behaviour is still described by the overconsolidated model. The modulus number χ and exponent α take their overconsolidated values, χ_{oc} and α_{oc}.

When the degree of overconsolidation is low, so that n is not much greater than 1, the exponent α probably does not change much between the normally compressed and the overconsolidated states, $\alpha_{oc} \approx \alpha_{nc}$, and the principal difference is in the modulus number, $\chi_{oc} > \chi_{nc}$. Thus for soft clays, normally compressed or lightly overconsolidated – $n < 2$, say – a reasonable initial estimate might be $\alpha = 1$ and $\chi_{oc} \approx 5\chi_{nc}$. For heavily overconsolidated clays, from which large historical depths of overlying soil, rock or ice have been removed and $n > 10$, say, it might be more appropriate to take a much lower value of $\alpha \approx 0.1$ and much higher value of $\chi \approx 1000$, for example. We might reasonably ask what would happen if such a heavily overconsolidated clay were subjected to such a high increase in stress that it actually became

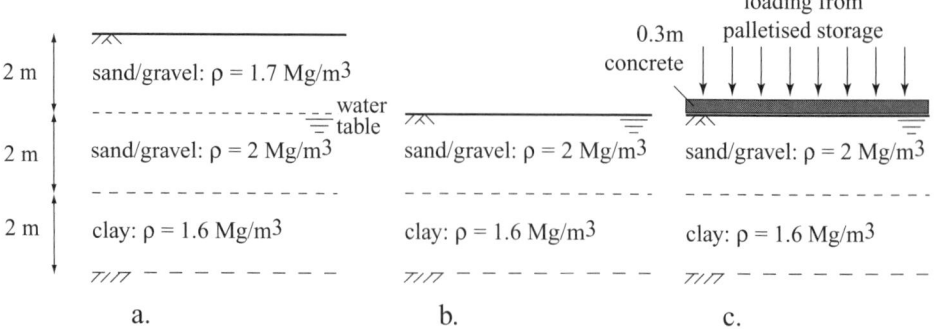

Figure 4.18. Site conditions (a) before and (b) after excavation of top 2 m of sand/gravel; and (c) after construction of concrete slab and application of load from palletised storage.

normally compressed once again. Our (weak) response has to be to admit that it is not intended that our simple model should be used for such extreme circumstances.

4.7.1 Worked examples: Overconsolidation

At a particular site, the ground conditions consist of 4 m of sands and gravels overlying 2 m of soft, normally compressed clay (Fig. 4.18a). The water table is at a depth of 2 m. The sands and gravels have a density of 2 Mg/m³ below and 1.7 Mg/m³ above the water table. The clay has density 1.6 Mg/m³ and is characterised by stiffness properties: exponent $\alpha_{nc} = \alpha_{oc} = 1$, modulus number $\chi_{nc} = 10$, $\chi_{oc} = 50$.

1. The sands and gravels are excavated to a depth 2 m over an area of large extent (Fig. 4.18b), so that we can consider the problem one-dimensional. What will be the eventual heave of the ground? The stiffness of the sands and gravels is so high that their contribution to the heave can be ignored.

    Let us treat the clay layer as a single uniform layer, characterised by the conditions at its mid-depth, 5 m below the original ground surface. The total stress at this depth, before the removal of the top 2 m of sands and gravels, is: $\sigma_{z1} = \Sigma \rho g \Delta z = (2 \times 1.7 + 2 \times 2 + 1 \times 1.6) \times 9.81 = 88.29$ kPa. The pore pressure at this depth, 3 m below the water table, is $u = 3 \times 9.81 = 29.43$ kPa, and hence the initial effective stress is $\sigma'_{z1} = \sigma_{z1} - u = 58.86$ kPa.

    The eventual effect of the excavation is to reduce the total stress and effective stress by $\Delta\sigma_z = \Delta\sigma'_z = -2 \times 1.7 \times 9.81 = -33.35$ kPa so that the final effective stress is $\sigma'_{z2} = 58.86 - 33.35 = 25.52$ kPa.

    Because $\alpha = 1$, we have to use (4.26) to calculate the strain in the clay. The clay is being unloaded from a normally compressed state and therefore becomes overconsolidated. We need to make use of the overconsolidated value of the modulus number χ_{oc}:

$$\varepsilon_z = \frac{1}{\chi_{oc}} \ln \frac{\sigma'_{z2}}{\sigma'_{z1}} = \frac{1}{50} \ln \frac{25.52}{58.86} = -0.0167$$

4.7 Overconsolidation

or a strain of −1.67%. The negative sign indicates that the clay is heaving or expanding in thickness because of the reduction of effective stress. This is the strain; the heave is the product of the strain and the thickness of the layer of clay: $\varsigma = \varepsilon_z \Delta z = -0.0167 \times 2 = -0.033$ m.

2. The excavation was part of the initial stage of construction of a large warehouse which will consist roughly of a concrete floor 0.3 m thick, of density 2.4 Mg/m³, and subsequent palletised stored contents (Fig. 4.18c). What will be the eventual settlement of the site – which again means just the clay – if the stored materials generate a pressure of (a) 25 kPa or (b) 40 kPa?

The concrete floor generates a surface pressure of $0.3 \times 2.4 \times 9.81 = 7.06$ kPa. With a loading of 25 kPa from the stored materials, the total increase in applied vertical stress is $7.06 + 25 = 32.06$ kPa, which is just a little less than the stress removed by the excavation of the top 2 m of sands and gravels (33.35 kPa). The whole of this stress increase can thus be absorbed within the overconsolidated response of the clay. The vertical effective stress changes from $\sigma'_{z2} = 25.52$ kPa to $\sigma'_{z3} = 25.52 + 32.06 = 57.58$ kPa and the resulting vertical strain is:

$$\varepsilon_z = \frac{1}{\chi_{oc}} \ln \frac{\sigma'_{z3}}{\sigma'_{z2}} = \frac{1}{50} \ln \frac{57.58}{25.52} = 0.0163$$

or a strain of 1.63%. The settlement is the product of the strain and the thickness of the layer of clay: $\varsigma = \varepsilon_z \Delta z = 0.0163 \times 2 = 0.032$ m. The nett effect of the construction activities is to produce a negligible heave with a magnitude of about 1 mm.

With an applied operational loading of 40 kPa, we can see that we will be taking our clay back into the normally compressed region. The total increase in effective stress is the sum of the stress generated by the concrete slab and this applied loading: $7.06 + 40 = 47.06$ kPa. We have to divide this increase in vertical effective stress into two parts: the first part, 33.35 kPa, exactly matches the stress removed through excavation of the top 2 m of sands and gravels. This will exactly recover in settlement the heave of 0.033 m that occurred as a result of the excavation. The remainder, $47.06 - 33.35 = 13.71$ kPa, is compressing the clay in the normally compressed, less stiff region. The final vertical effective stress is $\sigma'_{z4} = 25.52 + 47.06 = 72.58$ kPa and we can apply (4.26) to calculate the strain, using the normally compressed value of the modulus number χ_{nc}:

$$\varepsilon_z = \frac{1}{\chi_{nc}} \ln \frac{\sigma'_{z4}}{\sigma'_{z1}} = \frac{1}{10} \ln \frac{72.58}{58.86} = 0.0209$$

or a strain of 2.09%. The total strain resulting from the use of the warehouse with this level of loading is therefore $0.0167 + 0.0209 = 0.0376$ or 3.76%, and the settlement is 0.075 m. The nett effect of the construction activities is to produce a settlement of about 0.042 m.

It is evident that the settlement increases rapidly once the loading takes the clay into a normally compressed state of stress. There is obvious advantage in trying to

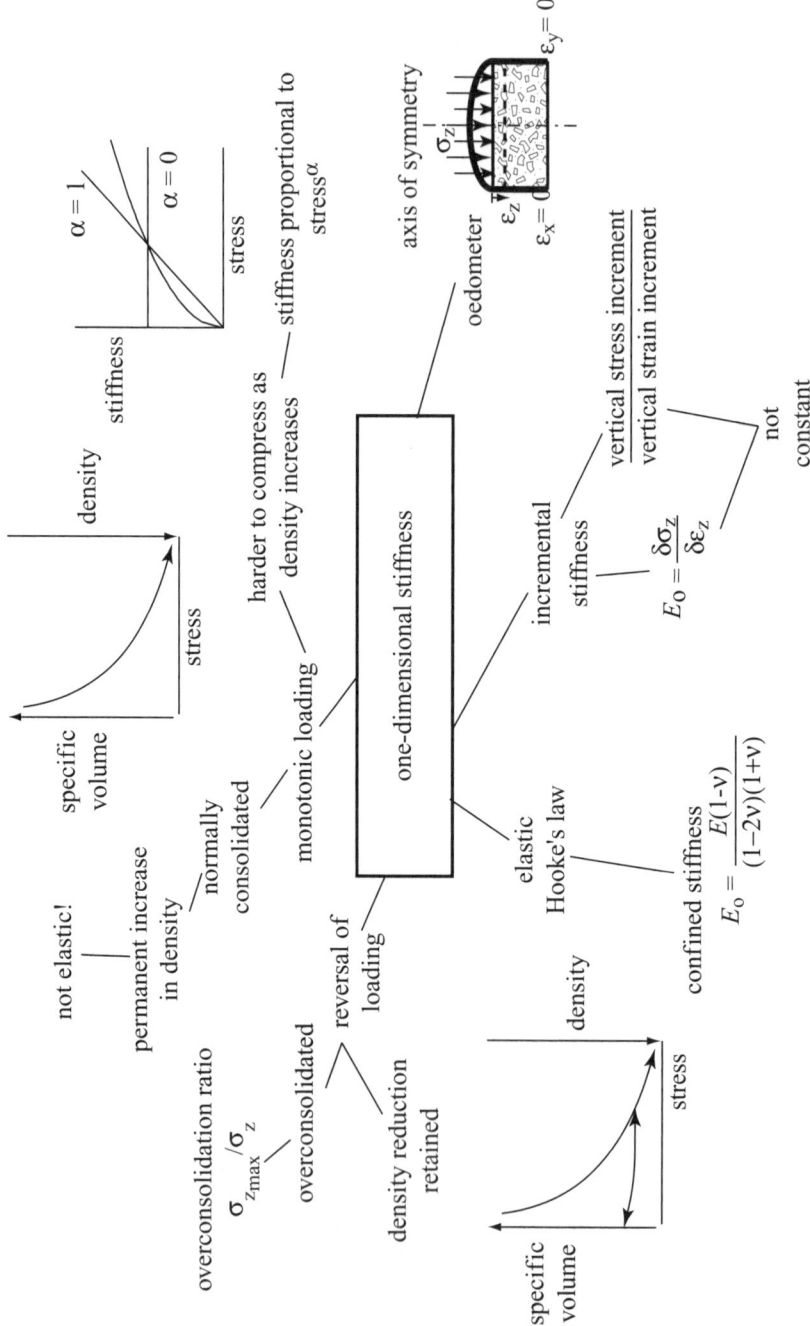

Figure 4.19. Mind map: one-dimensional stiffness.

4.8 Summary

design a balanced excavation in which the stress removed through excavation more or less matches the eventual working loading that the ground will have to sustain.

4.8 Summary

Here is a concise list of the key messages from this chapter, which are also encapsulated in the mind map (Fig. 4.19).

1. Stiffness is a general term describing the link between changes in strain and changes in stress.
2. Linear elasticity is a particular (simple) form of stiffness model described, for example, by Hooke's Law.
3. As soils are compressed one-dimensionally in an oedometer, the density increases and the stiffness also increases.
4. A power-law expression links one-dimensional stiffness to the current stress with an exponent α and a modulus number χ. Various different types of response can be described using different values of α: $\alpha = 0$ implies constant stiffness or linear elasticity.
5. There is some correlation between the values of χ and soil density.
6. The compression of soils is often irreversible: unloading is accompanied by smaller changes in volume and higher stiffnesses and permanent increase in density.
7. Overconsolidation ratio (ocr) describes the ratio of maximum past vertical stress to current vertical stress.
8. Normally compressed soils have an overconsolidation ratio ocr $= 1$.

4.9 Exercises: Stiffness

1. Using the general expression for incremental one-dimensional stiffness E_o with reference stress $\sigma_{ref} = 100$ kPa calculate the stiffnesses for the geotechnical materials in Table 4.2 at a vertical effective stress $\sigma_z = 150$ kPa.

   Table 4.2. *Stiffness properties for Question 1.*

	α	χ
rock	0	10^5
glacial till	0.2	10^3
sand	0.5	10^2
clay	1	10

2. Use the stiffnesses calculated in Question 1 to calculate the strain generated when the vertical effective stress increases from 150 kPa to 165 kPa. (Hint: Assume that the stress increment is sufficiently small for the stiffness to be regarded as constant over the increment.)

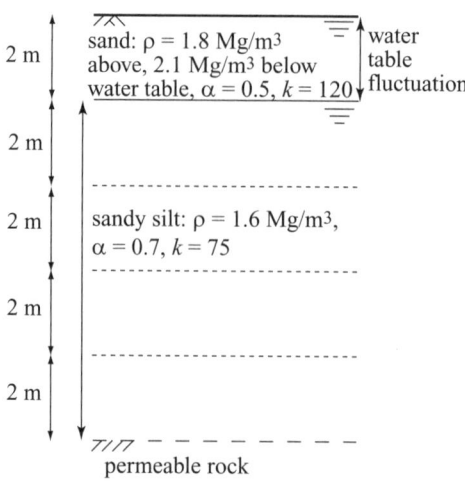

Figure 4.20. Site conditions for Question 6: water table varies between ground surface and depth of 2 m.

3. For the four materials in Question 1, calculate the strain generated when the vertical effective stress increases from 150 kPa to 300 kPa. Compare the average stiffnesses with the stiffnesses calculated in Question 1.
4. The average vertical effective stress in a layer of clay of thickness 5 m, with properties taken from Table 4.2, increases from 150 kPa to 300 kPa. Calculate the corresponding change in thickness of the layer.
5. The average vertical effective stress in a layer of sand of thickness 4 m, with properties taken from Table 4.2, increases from 150 kPa to 400 kPa. Calculate the corresponding vertical strain and the change in thickness of the layer.
6. The soils at a site consist of 2 m of sand over 8 m of sandy silt over permeable rock (Fig. 4.20). The water table is initially at the ground surface but, as a result of pumping for water supply, falls to a depth of 2 m.
   (a) The density of the sand is 2.1 Mg/m³ below and 1.8 Mg/m³ above the water table. The density of the silt is 1.6 Mg/m³. Divide the sandy silt into layers of thickness 2 m, as shown in Fig. 4.20, and calculate the effective vertical stress at the centre of each layer of soil before and after the water table is lowered.
   (b) The stiffness of the sand is characterised by $\chi = 120$, $\alpha = 0.5$ and the silt by $\chi = 75$, $\alpha = 0.7$. Calculate the strain that occurs in each soil layer as a result of the lowering of the water table. Hence, calculate the settlement at the ground surface.
7. At a coastal site, the ground conditions consist of 8 m of silty sand overlying 2 m of gravel and then rock (Fig. 4.21). The density of the silty sand is 1.6 Mg/m³ above the water table and 2.0 Mg/m³ below the water table. The density of the gravel is 2.1 Mg/m³. The water table varies regularly between 1 m and 4 m below the ground surface. What are the maximum and minimum vertical effective stresses at a depth of 4 m, and what is the corresponding range of overconsolidation ratio?

4.9 Exercises: Stiffness

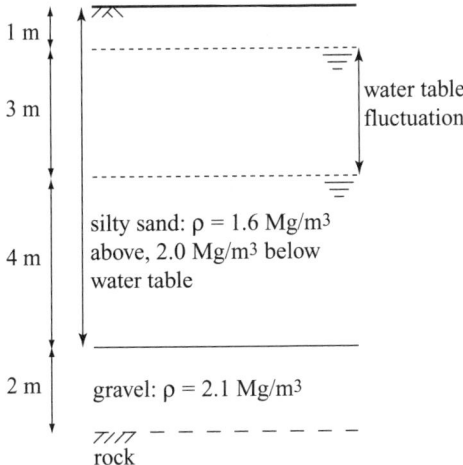

Figure 4.21. Site conditions for Question 7: water table varies between depths of 1 m and 4 m.

Making reasonable assumptions, choose values of stiffness exponent α and modulus number χ (Fig. 4.11) for the silty sand and estimate the heave and settlement that occur each time the water table varies between its extreme levels.

8. As a result of a change in the economic climate, the design for the warehouse in Section 4.7.1 (Fig. 4.18) has been changed. The sands and gravels will now only be excavated to a depth of 1 m. Calculate the heave that will result from this excavation and the eventual settlement under working storage loads of 25 kPa and 40 kPa. The thickness of the concrete slab remains unchanged at 0.3 m.
9. The results of an oedometer test on reconstituted Gault clay are given in Table 4.3. The specific volume of the clay at the end of the test was found to be 1.979. Estimate the value of modulus number χ and stiffness exponent α for this clay in its normally consolidated and overconsolidated states.

Table 4.3. *Oedometer test on reconstituted Gault clay (Question 9).*

Vertical stress kPa	Sample height mm
0	19.05
50	17.95
100	16.77
200	15.56
400	14.49
100	14.96
50	15.28

5 Seepage

5.1 Introduction

Water is a particular source of geotechnical problems (Figs 5.1, 5.2). It is no coincidence that landslides frequently occur during or after periods of heavy rainfall. We introduced some of the basic principles of hydrostatics in Chapter 2 and used the idea of a water table to calculate pore water pressures in the ground and thus convert total stresses (equilibrium) to effective stresses (which control mechanical response of soils). Hydrostatics is of course concerned with water at rest – here we will allow the water to move through the soil (but not very fast) and introduce principles of one-dimensional seepage.

5.2 Total head: Bernoulli's equation

There are some basic building blocks that will assist our study of seepage. One is Bernoulli's equation, which describes the steady flow of an incompressible fluid along a streamline, or through a frictionless tube. This is obviously a somewhat idealised situation but the assumption of incompressibility is certainly reasonable for the flow of water under the pressures that are likely to occur in most civil engineering systems. The flow rates in soils will generally be slow.

A reference diagram is shown in Fig. 5.3 for an element of water of density ρ_w of cross-section A which is flowing vertically with velocity v and with pressures u acting at its base, at level z above some reference datum, and $u + \delta u$ at its top, at level $z + \delta z$ above the same datum.[1] The mass of the element is $\rho_w A \delta z$. There is a downward body force on the element due to gravity (mass × acceleration), $\rho_w g A \delta z$. As a result of the forces acting on the element due to the fluid pressure and the gravitational body force, we would expect from Newton's laws of motion that the

---

[1] In Chapter 2, we found it convenient to take a sign convention for vertical positions in which we started at the ground surface and measured depth z positive downwards. However, in analysing flow of water, it is really more helpful to measure height *above* some reference datum so that the coordinate z will be positive upwards. Beware!

5.2 Total head: Bernoulli's equation

Figure 5.1. Small slope failure triggered by a burst water main – no injuries (Tsz Wan Shan, Hong Kong).

fluid element would accelerate:

$$uA - (u + \delta u)A - \rho_w g A \delta z = \rho_w A \delta z \frac{dv}{dt} \qquad (5.1)$$

Acceleration a is the rate of change of velocity v with time t, so it is natural to write $a = dv/dt$ (Fig. 5.4a), but we can also look at acceleration in a different way. Imagine that we have an object moving along a track in the z direction. At one moment it is at position z, the next moment it is at position $z + \delta z$. As it moves from

Figure 5.2. Slope failure in Hong Kong triggered by heavy rainfall (Po Shan Road, 18 June 1972) (*http://hkss.cedd.gov.hk/hkss/eng/photo_gallery/Landslide/index_1.htm*).

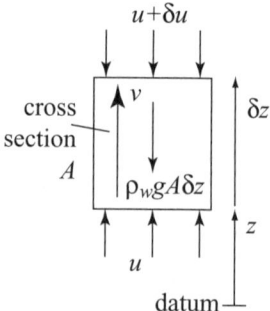

Figure 5.3. Element of water subject to self weight and gradient of pore pressure.

z to $z + \delta z$, the velocity changes from v to $v + \delta v$ (Fig. 5.4b). The velocity evidently changes with both position and time: we can write $v = v(t)$ or $v = v(z)$, but of course the two are not independent. The time taken to move from position z to position $z + \delta z$ is $\delta t = \delta z/v$ (neglecting the small change of velocity that occurs during the move). The rate of change of velocity with time – the acceleration a – is then:

$$a = \frac{\delta v}{\delta t} = \frac{\delta v}{\delta z/v} = v\frac{\delta v}{\delta z} \tag{5.2}$$

and in the limit, as the changes in position and time become infinitesimally small:

$$a = \frac{dv}{dt} = v\frac{dv}{dz} \tag{5.3}$$

and it is often convenient to make use of this alternative differential form of the acceleration.[2]

Thus (5.1) becomes:

$$\delta u + \rho_w g \delta z + \rho_w v \frac{dv}{dz}\delta z = 0 \tag{5.4}$$

The pore pressure u varies with position, so we can write $u = u(z)$ and:

$$\delta u = \frac{du}{dz}\delta z \tag{5.5}$$

We can then write (5.4) as:

$$\left(\frac{du}{dz} + \rho_w g + \rho_w v\frac{dv}{dz}\right)\delta z = 0 \tag{5.6}$$

---

[2] We could deduce the same result by using the chain rule for differentiation of the function for the velocity $v = v(z)$:

$$\frac{dv}{dt} = \frac{dv}{dz}\frac{dz}{dt} = v\frac{dv}{dz}$$

noting that $v = dz/dt$.

5.2 Total head: Bernoulli's equation

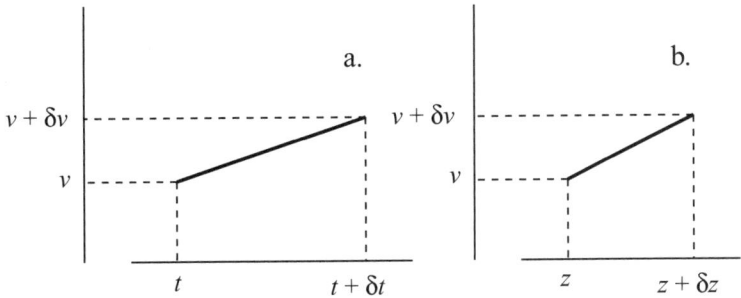

Figure 5.4. Change of velocity v with (a) time t and (b) position z.

and we can integrate this to find the changes that occur as we move (one-dimensionally) through space:

$$\int \left(\frac{du}{dz} + \rho_w g + \rho_w v \frac{dv}{dz} \right) dz = \int \frac{du}{dz} dz + \int \rho_w g \, dz + \int \rho_w v \frac{dv}{dz} dz = \int 0 \, dz \quad (5.7)$$

so that:

$$u + \rho_w g z + \frac{1}{2} \rho_w v^2 = \text{constant} \quad (5.8)$$

This is Bernoulli's equation, which indicates that, in the absence of frictional losses, the sum of pressure u, plus potential energy $\rho_w g z$, plus kinetic energy $\frac{1}{2}\rho_w v^2$ is constant along a flow path. The potential energy is the work done against gravity in lifting an element of water to its current elevation z above the chosen datum: hydroelectric power is all about converting potential energy into kinetic energy to drive electricity-generating turbines. Although we have approached this result from consideration of a one-dimensional system (Fig. 5.3), the result is general – kinetic energy is a scalar quantity.

One very visual way of measuring fluid pressure is to place a standpipe in the fluid and observe the level to which the fluid rises (Fig. 5.5). The height h of the fluid surface above the measuring point A then indicates that the pressure is $u = \rho_w g h$ as we saw in Sections 2.7 and 2.8 (see also Fig. 2.15). The height h is the pressure *head*

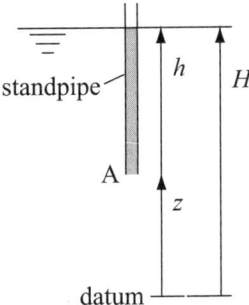

Figure 5.5. Standpipe to measure water pressure head h; elevation head z, total head H and reference datum.

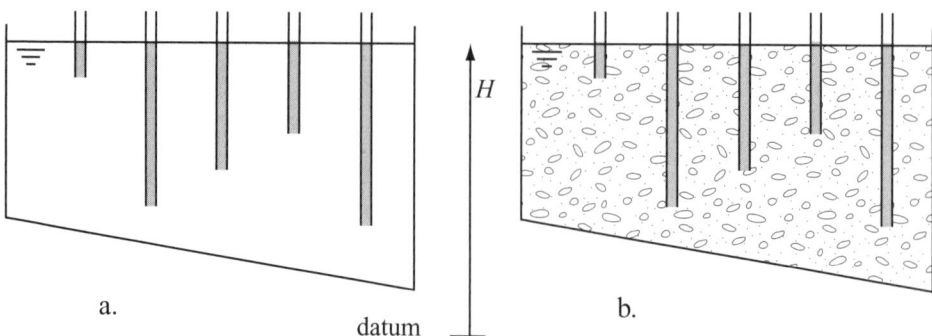

Figure 5.6. (a) Total head in pool of water; (b) total head in pool of water and soil.

at that point. We can also define a velocity head, h_v, by dividing the kinetic energy term by the unit weight of water $\rho_w g$, so that Bernoulli's equation (5.8) becomes:

$$H = \frac{u}{\rho_w g} + z + \frac{1}{2}\frac{\rho_w v^2}{\rho_w g} = h + z + h_v = \text{constant} \qquad (5.9)$$

and we call H the *total* head in the fluid, distinguishing it from the *pressure* head h and the *elevation* head z. Elevation head obviously requires a reference datum and so also does total head – and the reference datum must be the same for both elevation and total heads. On the other hand, pressure head h is a difference between these two and is independent of the choice of datum (Fig. 5.5). For stationary fluid, the velocity and kinetic energy are zero ($v = h_v = 0$) and, in the absence of flow, the sum of pressure and potential energy (or pressure head and elevation head) is constant:

$$H = h + z = \text{constant} \qquad (5.10)$$

If we think once again of a swimming pool full of water (Fig. 5.6a) then, ignoring convection currents resulting from temperature differences, or currents induced by energetic swimmers (it is that time of early morning before the pool is open), we can understand that the total head is the same everywhere and it is equal to the level of the surface of the pool. The water in any standpipe that we place or suspend in the pool will rise to the level of the water surface at height H above the chosen datum. If we fill the pool with soil, displacing some of the water, and allow everything to settle down (the pool has lost its attraction to swimmers) the level of the water surface will be unchanged and the idea of the constancy of total head remains intuitively reasonable (Fig. 5.6b). Typically, our standpipe will now be fitted with a filter at its base (Fig. 5.7) so that the water can enter freely but the soil is prevented from entering the tube: such a device for measuring pore water pressure in a soil is called a *piezometer*. The water will still rise to the same level H above the datum in standpipes located anywhere in the soil-filled pool.

Now let us think of two pools or tanks filled with water and connected by several tubes (Fig. 5.8). If the level of the water surface in the two tanks is the same (Fig. 5.8a) then there will be no flow – irrespective of the angle of the tube. At end A of

5.2 Total head: Bernoulli's equation

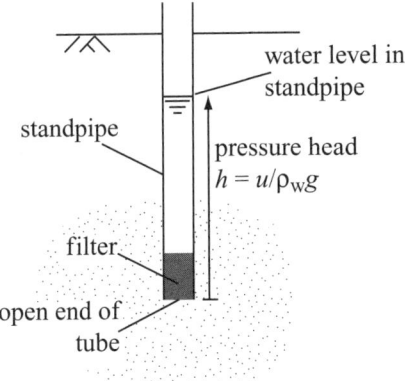

Figure 5.7. Standpipe piezometer for measuring pore water pressure in the ground.

a typical tube (Fig. 5.8b), at elevation z_1 relative to the datum, the pressure head is h_1; at end B of the tube, at elevation z_2 relative to the datum, the pressure head is h_2. However, as indicated in Fig. 5.8b, the total head, $H = h_1 + z_1 = h_2 + z_2$, is the same at both ends A and B of the tube. Even though the pressure head may be very different at the two ends of the tube (Fig. 5.8b), the differences in elevation head compensate so that the total head is the same.

This conclusion is independent of the size (bore) of the connecting tube, so let us make the diameter of the tube progressively smaller (Fig. 5.9). Provided the total head in each tank is the same, there will be no flow through the tube (Fig. 5.9a). But suppose that we now lower the water level in one of the tanks (Fig. 5.9b). Water will flow from the tank with the higher water level (higher total head) to the tank with the lower water level (lower total head) – and again the direction of flow will be independent of the orientation of the connecting tube. For a large diameter of connecting tube, the flow rate would be as fast as we could top up the tank with the higher total head, and the velocity head in Bernoulli's equation (5.8, 5.9) might

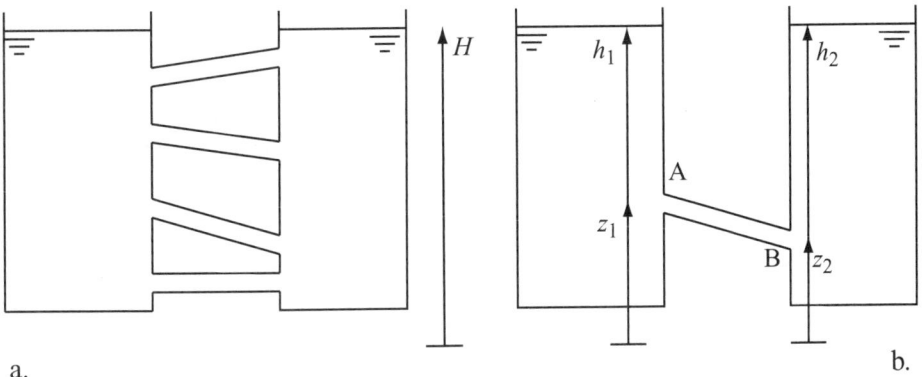

Figure 5.8. (a) Two tanks of water connected by tubes at different inclinations; (b) different pressure heads but equal total heads.

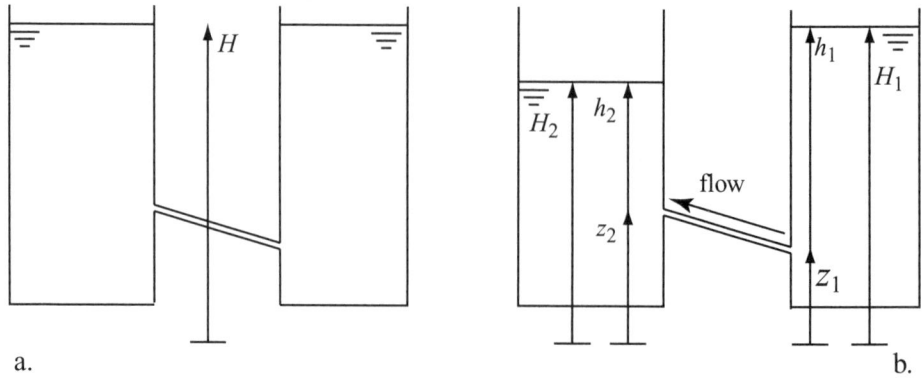

Figure 5.9. (a) Two tanks of water connected by narrow tube; (b) flow induced by difference in total head across ends of tube.

be significant. But if we make the tube quite small, then we expect that viscous effects and the frictional resistance of the walls of the tube will tend to impede the flow: recall that we said that Bernoulli's equation applied in the absence of frictional losses. Flow will occur but the velocity of flow will be very small.

5.3 Poiseuille's equation

A second building block is then Poiseuille's equation, which describes the flow in narrow tubes. Once again some assumptions are required. We assume that the flow in the tube is laminar, which means that it is smooth and not turbulent. We assume that the water (or other flowing fluid) is incompressible and has a viscosity η. Viscosity describes the way in which shear stresses are generated whenever there is a gradient of velocity (recall Section 3.6.2). Thus if one (infinitesimally thick) layer of water is moving slightly faster than the neighbouring layer of water, then the slower layer will try to hold back the faster layer with a shear stress (Fig. 5.10):

$$\tau = \eta \frac{dv}{dz} \tag{5.11}$$

In our tube, of radius r_o, (Fig. 5.11a), the variations of velocity will occur with the radius: the profile of velocity will be expected to be symmetrical about the centre of the tube (Fig. 5.11b). The wall of the tube is sufficiently rough to ensure that the

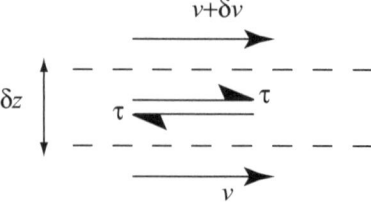

Figure 5.10. Viscous shear stress τ between faster and slower layers of liquid.

5.3 Poiseuille's equation

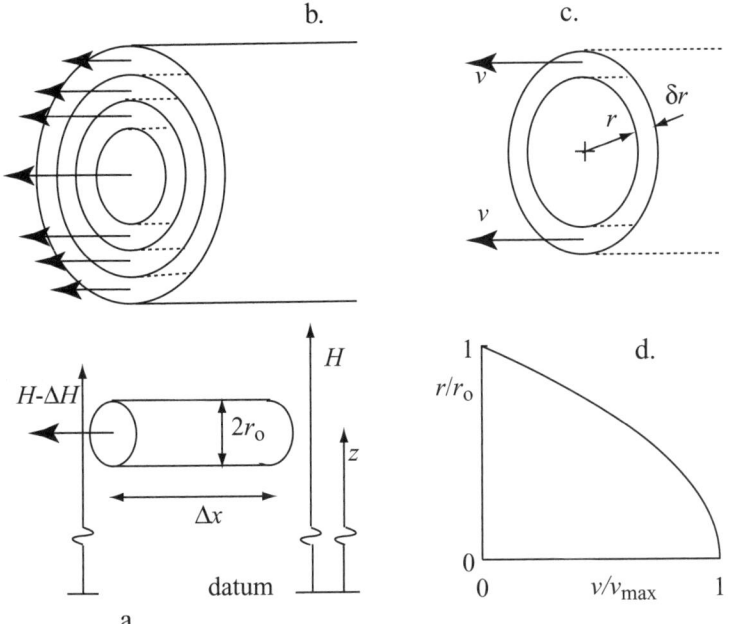

Figure 5.11. (a) Flow of water through tube of length Δx as a result of a difference of total head ΔH; (b), (c) elemental rings of water; (d) parabolic radial variation of velocity.

velocity at $r = r_o$ will be zero. Thus the velocity is expected to vary smoothly from the wall of the tube, where it is zero, to the centre of the tube, where it is a maximum.

We finally assume that conditions are steady and that the water is not accelerating through the tube. This seems consistent with our ideas of conservation of mass: the mass of water flowing into the tube must exactly balance the mass of water flowing out of the tube. The water is more or less incompressible, so, if the cross-sectional area of the tube is constant, the velocity must be constant, too.

Let us consider an annular ring of water (Fig. 5.11c) at radius r and with thickness δr, where the velocity is v. The ring has length Δx (Fig. 5.11a) and there is a pressure differential Δu in the water between the ends of this annular element: this pressure differential will be the same at all radii. Our discussion of Bernoulli's equation and the significance of total head tells us that in fact it is the differential of total head ΔH (Fig. 5.11a) rather than pressure that will drive the flow, even in a narrow tube. We have drawn the tube horizontal for convenience, located at height $z \gg r_o$ above our datum, in Fig. 5.11 but the usual sort of thought experiment placing a narrow tube in a swimming pool at different inclinations (Fig. 5.12) confirms that there will be no flow through the tube simply because there is a pressure difference between the ends if this pressure difference is the result purely of differences in elevation of the two ends of the tube: $\Delta u = -\rho_w g \Delta z$, $\Delta H = 0$.

Since there is no acceleration, there must be a balance between the force provided by the differential total head ΔH across the element and the force provided by the shear stress resulting from the radial gradient of velocity. The water in the

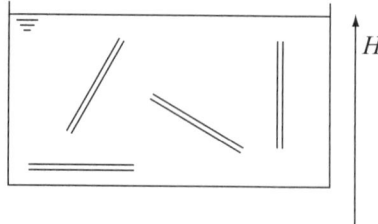

Figure 5.12. Narrow tubes at different inclinations in a pool of water.

annular element is being pulled forward by the water nearer the centre of the tube and pulled back by the water nearer the wall of the tube. The radial gradient of velocity is in fact negative but, in writing our equation of equilibrium, we simply use the velocity and its radial derivatives and expect negative values to emerge from our analysis.

The pressure in the water at input to the tube is $u_i = \rho_w g(H - z)$, providing a force $\rho_w g(H - z)2\pi r \delta r$ on the annular element of water between radii r and $r + \delta r$. We assume that $r \ll (H - z)$ so that the pressure u is essentially constant across the whole section of the tube in Fig. 5.11a. The pressure in the water at the outlet of the tube is $u_o = u_i - \Delta u = \rho_w g(H - \Delta H - z)$, providing a force on the other end of our annular element of $\rho_w g(H - \Delta H - z)2\pi r \delta r$. The out-of-balance force on the annular element, tending to push the water through the tube against the viscous drag of the neighbouring elements, is $\rho_w g \Delta H 2\pi r \delta r$. The viscous drag from the water on the inside of the annular element is a shear stress equal to the viscosity multiplied by the velocity gradient, dv/dr, acting over an area $2\pi r \Delta x$. The viscous drag from the water on the outside of the annular element at radius $r + \delta r$ is a shear stress equal to the viscosity multiplied by the velocity gradient at that radius, $dv/dr + [d(dv/dr)dr]\delta r$, acting over the slightly larger area $2\pi(r + \delta r)\Delta x$. Both the area of the surface of the annular element and the velocity gradient change with radius. Putting these force components into an equation of equilibrium:

$$\rho_w g \Delta H 2\pi r \delta r = \eta \frac{dv}{dr} 2\pi r \Delta x - \eta \left[\frac{dv}{dr} + \frac{d}{dr}\left(\frac{dv}{dr}\right) \delta r \right] 2\pi(r + \delta r) \Delta x \quad (5.12)$$

Expanding the products this gives:

$$\rho_w g \Delta H r \delta r = -\eta \left[r \frac{d}{dr}\left(\frac{dv}{dr}\right) \delta r + \frac{dv}{dr} \delta r + \frac{d}{dr}\left(\frac{dv}{dr}\right)(\delta r)^2 \right] \Delta x \quad (5.13)$$

Ignoring second order quantities – the terms containing the product $(\delta r)^2$ are certainly negligible in comparison with terms which only contain the first order small quantities δr – this becomes:

$$\frac{d^2 v}{dr^2} + \frac{1}{r}\frac{dv}{dr} = \frac{1}{r}\frac{d}{dr}\left(r \frac{dv}{dr}\right) = -\frac{\rho_w g}{\eta} \frac{\Delta H}{\Delta x} \quad (5.14)$$

and the right-hand side of this equation is constant, independent of radius.

5.4 Permeability

Integrating this:

$$r\frac{dv}{dr} = -\frac{\rho_w g}{2\eta}\frac{\Delta H}{\Delta x}r^2 + J_1 \qquad (5.15)$$

where J_1 is a constant of integration. However, from symmetry the gradient of radial velocity must be zero at the centre of the tube, $r = 0$, and hence $J_1 = 0$ and:

$$\frac{dv}{dr} = -\frac{\rho_w g}{2\eta}\frac{\Delta H}{\Delta x}r \qquad (5.16)$$

and the (linear) variation of shear stress with radius is:

$$\tau = \eta\frac{dv}{dr} = -\frac{\rho_w g}{2}\frac{\Delta H}{\Delta x}r \qquad (5.17)$$

Integrating (5.16):

$$v = -\frac{\rho_w g}{4\eta}\frac{\Delta H}{\Delta x}r^2 + J_2 \qquad (5.18)$$

The velocity is zero at the wall of the tube $r = r_o$, and hence the integration constant J_2 is:

$$J_2 = \frac{\rho_w g}{4\eta}\frac{\Delta H}{\Delta x}r_o^2 \qquad (5.19)$$

and we have recovered a parabolic velocity profile (Fig. 5.11d):

$$v = \frac{\rho_w g}{4\eta}\frac{\Delta H}{\Delta x}(r_o^2 - r^2) \qquad (5.20)$$

The volumetric flow rate through the tube is:

$$q = \int_0^{r_o} 2\pi r v \, dr = \frac{\rho_w g \pi}{2\eta}\frac{\Delta H}{\Delta x}\int_0^{r_o} r(r_o^2 - r^2) dr = \frac{\rho_w g \pi r_o^4}{8\eta}\frac{\Delta H}{\Delta x} \qquad (5.21)$$

so that the average velocity $\bar{v}$ through the tube is:

$$\bar{v} = \frac{q}{\pi r_o^2} = \frac{\rho_w g r_o^2}{8\eta}\frac{\Delta H}{\Delta x} \qquad (5.22)$$

These are alternative ways of presenting Poiseuille's equation.

5.4 Permeability

What is the relevance of Poiseuille's equation to the flow of water through soils? Soils obviously do not consist of straight cylindrical tubes. In fact the pore spaces through which water has to flow are geometrically significantly more complex than cylinders (Fig. 3.1). There is no such thing as a "cross section" and the path that the water has to follow is extremely tortuous. However, what Poiseuille does tell us is that the ease with which water flows through an opening will be proportional to the square of the size of that opening. We have seen (Fig. 3.8) that the size of soil particles varies over a range of many orders of magnitude – gravel particles are typically some 10^5 times larger than clay particles – and we might expect that the

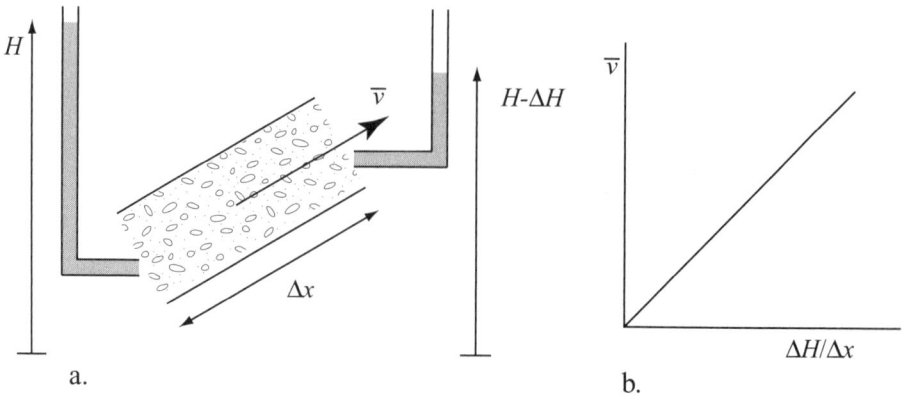

Figure 5.13. Darcy's Law.

typical dimensions of the spaces between the particles would be of the same order as the particle sizes. So without any further investigation we might suppose that the ease with which water can flow through a gravel would be 10^{10} times greater than through a clay.

Henry Darcy was responsible for the water supply to the town of Dijon and performed experiments to discover how water would flow through sand filters.[3] Flow of water through soils is usually described using Darcy's Law (Fig. 5.13). We write:

$$\bar{v} = k\frac{\Delta H}{\Delta x} = ki \qquad (5.23)$$

where $\bar{v}$ is the flow velocity averaged over the cross-section of the soil, k is the coefficient of permeability, and $i = \Delta H/\Delta x$ is known as the *hydraulic gradient* and is a non-dimensional way of describing the gradient of total head within the soil. Permeability k thus has dimensions of velocity: length/time.

The similarity between Darcy's Law and Poiseuille's equation is clear – and the "permeability" of the narrow tubes comparing (5.22) and (5.23) is:

$$k_{tube} = \frac{\rho_w g r_o^2}{8\eta} \qquad (5.24)$$

An alternative description of permeability uses an absolute or specific permeability K:

$$K = \frac{k\eta}{\rho_w g} \qquad (5.25)$$

and it is found that, for a given soil, K is more or less a soil constant (provided the void ratio or the tube radius is not changed) which is independent of the actual

---

[3] Darcy, H. (1856) *Les fontaines publiques de la ville de Dijon: exposition et application des principes à suivre et des formules à employer dans les questions de distribution d'eau; ouvrage terminé par un appendice relatif aux fournitures d'eau de plusieurs villes au filtrage des eaux et à la fabrication des tuyaux de fonte, de plomb, de tole et de bitume.* Paris, Victor Dalmont, Éditeur, Successeur de Carilian-Gœury et Victor Dalmont, Libraire des Corps Impériaux des Ponts et Chaussées et des Mines, Quai des Augustins, 49.

5.4 Permeability

Table 5.1. *Typical values of soil permeability.*

	k m/s
clean gravel	$>10^{-2}$
sands and gravel mixtures	10^{-4}–10^{-2}
fine sands and silts	10^{-9}–10^{-4}
clays	$<10^{-9}$

permeating fluid (which might have different density and viscosity) and external conditions such as temperature which will affect the viscosity. The specific permeability K has units of length squared and is also expressed in darcys where 1 darcy = 0.987×10^{-14} m^2. Comparing (5.25) with (5.24) we can see that the specific permeability of Poiseuille's narrow tube is $K_{tube} = r_o^2/8$, which reinforces the assertion that K should be a more fundamental property than k. However, in soil mechanics we generally use the permeability k.

Typical values of permeability k of soils are shown in Table 5.1, and these confirm our expectation from Poiseuille's equation that the values of permeability will cover a very wide range. The permeability of gravels can be as high as 1 m/s and the permeability of clays can be as low as 10^{-12} m/s. If we want to find a soil material to use as a drain through which the water will pass without hindrance, then we will use a coarse sand or gravel; if we want to find a soil material which will act as a barrier to the passage of water – for example, to form the core of a dam (Figs 1.12, 1.13) or a containment for noxious waste – then we will use a clay.

Attentive readers will have noted that when we introduced Bernoulli's equation back in Section 5.2, we specifically ignored the kinetic energy or velocity head terms and dealt only with total head, elevation head and pressure head. Yet we are now calculating velocities of flow using gradients of total head with no reference at all to the kinetic energy term. We can make two points in our defence. First, we did say that Bernoulli's equation applied in the absence of frictional losses but the viscous resistance to the flow of water through the narrow pores of the soil (or through any narrow tube) will certainly provide important frictional losses in principle in the form of heat. Second, under typical geotechnical conditions a fast flow rate through the soil might be as high as 10 mm/s (0.01 m/s). The kinetic energy head for velocities of this magnitude is $h_v = \rho_w v^2 / 2\rho_w g = 0.01 \times 0.01/2 \times 9.81 = 5 \times 10^{-6}$ m, or 5 μm. We can reasonably assume this to be negligible.

Darcy's Law describes the average velocity through the cross-section of the soil, but we know that the space through which the water can actually flow is much less because the soil particles themselves occupy much of the space. If you inject some dye into the soil and monitor its transport, then you will see the seepage velocity v_s, which will be faster than the average velocity $\bar{v}$, the ratio of the two being linked with the void ratio e or porosity n of the soil. If we suppose that the ratio of void cross-sectional area (through which the water is flowing) to total cross-sectional area is the same as the ratio of volume of voids to total volume of the soil (an intuitively plausible but not inevitable assumption) we have:

$$v_s = \bar{v}\frac{1+e}{e} = \frac{\bar{v}}{n} \tag{5.26}$$

We can expect there to be differences in the way in which water flows through clays and sands simply resulting from the shapes of the particles: clay minerals have large specific surface (area of surface per mass of the mineral) and consequently large areas over which the water is close to the particle (Figs 3.1c, d). Besides, the structure of clay mineral molecules leads to the presence of some water molecules which are very strongly tied electrostatically to the clay mineral molecules. Sands and gravels tend to have much more rotund or sub-rotund particles – not actually round but not usually plate-like – and electrostatic effects are small.

For sands and gravels, some empirical formulae have been produced to link permeability with particle size and volumetric packing. Hazen's formula links permeability with the size of particles for which only 10% by mass of the soil is finer, d_{10} (see the particle size distributions in Figs 3.9, 3.10):

$$k = C_H d_{10}^2 \tag{5.27}$$

and this fits in with Poiseuille's expectation that the permeability will depend on the square of some typical dimension. Unfortunately, the range of values for the constant C_H is very large and the predictive benefit of the formula somewhat limited.

The formula proposed variously by Carman and Kozeny suggests that:

$$k = \frac{\rho_w g}{\eta} \frac{1}{C_{CK}} \frac{1}{S^2} \frac{e^3}{1+e} \tag{5.28}$$

where S is the surface to volume ratio for the soil particles, so that its reciprocal has the dimensions of length (matching Poiseuille again, compare (5.24) – much of the difference in the numerical values will come from the extreme tortuosity of the pore spaces and hence flow channels in soils) and C_{CK} is a soil constant. Carrier[4] suggests typically $C_{CK} \approx 5$. For spherical particles of radius r, $S = 3/r$ and with knowledge of the particle size distribution, an estimate of the overall effective value of S could be made (with the inclusion of a modifying shape factor if the particles were significantly non-spherical). The void ratio function $e^3/(1+e)$ in (5.28) is plotted in Fig. 5.14.

5.4.1 Darcy or Forchheimer?

Darcy developed his law linking flow and hydraulic gradient from the experimental study of the slow flow of water through filters. In engineering hydraulics, much use is made of a non-dimensional group of parameters called Reynolds' number to characterise different regimes of behaviour and, in particular, to identify the transition from laminar to turbulent flow. For flow of water through a porous medium such as

---

[4] Carrier, W.D. (2003) Goodbye, Hazen; Hello, Kozeny-Carman. *Journal of Geotechnical and Geoenvironmental Engineering*, ASCE **129**(11), 1054–1056.

5.4 Permeability

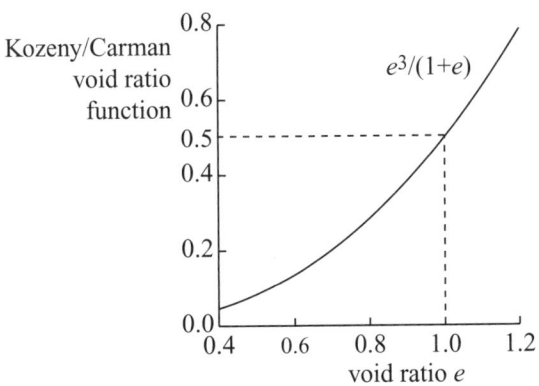

Figure 5.14. Carman and Kozeny relationship linking permeability and void ratio for a given soil.

a soil, we can define Reynolds' number R_e as:

$$R_e = \frac{\rho_w \bar{v} d_{50}}{\eta} \qquad (5.29)$$

where ρ_w and η are the density and the viscosity of the water, $\bar{v}$ is the velocity of flow, and d_{50} is a characteristic dimension which, for a granular material, can most conveniently be linked with particle size – for example, the size d_{50} for which half the soil particles are finer (Section 3.5 and Fig. 3.9). With this definition of Reynolds' number, it is found that the boundary between laminar and turbulent flow occurs for values of R_e of the order of 1 to 10.

Darcy's Law (5.23) provides a linear relationship between velocity and hydraulic gradient; Dupuit and Forchheimer[5] have proposed quadratic extensions of Darcy's Law to cope with faster, more turbulent, flows:

$$i = \frac{\bar{v}}{k} + \beta \frac{\bar{v}^2}{g} \qquad (5.30)$$

where β is possibly another soil property. This is known as the Forchheimer equation.

Most of the time we need not worry too much about the departure from linear Darcy flow. However, when we have coarse particles or rapid flows – for example, with the tide flowing in and out over a shingle beach (Fig. 3.1a) – then the non-linear expression may be important. Let us take some typical values: the boundary between medium and fine sand lies at 0.2 mm (Fig. 3.8); the density of water is $\rho_w = 1$ Mg/m^3; the viscosity of water at room temperature is around $\eta = 1$ mPa-s (Table 3.5). If we take $R_e = 1$ as the limit of applicability of Darcy's linear law, then this corresponds to a velocity of flow of:

$$\bar{v}_{crit} = \frac{R_e \eta}{\rho_w d_{50}} = \frac{1 \times 1 \times 10^{-3}}{1 \times 10^3 \times 0.0002} = 0.005 \text{ m/s} \qquad (5.31)$$

Thus a flow velocity of 5 mm/s is heading towards the boundary of the linear region.

---

[5] Dupuit, J. (1863) *Études théoriques et pratiques sur le mouvement des eaux.* Paris: Dunod. Forchheimer, P. (1901) *Wasserbewegung durch Boden.* Z. Ver. Deutsch Ing., **45**, 1782–1788.

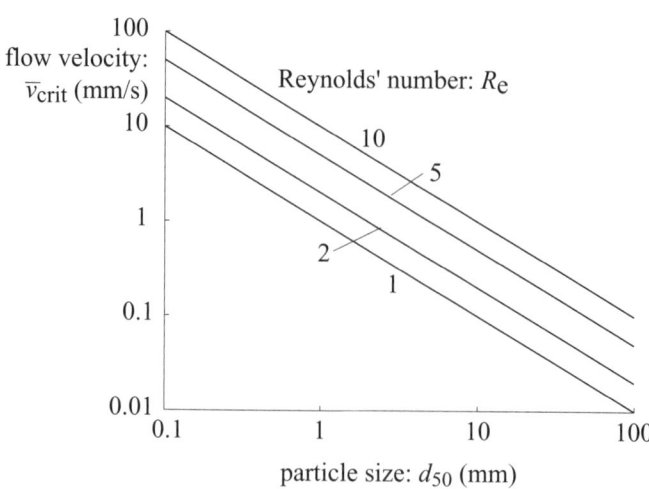

Figure 5.15. Boundary of applicability of linear Darcy's Law; flow velocity $\bar{v}$ as function of particle size d_{50} and Reynolds' number R_e.

The link between d_{50} and the critical velocity $\bar{v}_{crit}$ is shown in Fig. 5.15 for various values of Reynolds' number between 1 and 10. For shingle with particle size of around 50 mm, the boundary of turbulent flow for $R_e = 1$ comes with a velocity of only 20 μm/s!

5.5 Measurement of permeability

Techniques that are used for the measurement of permeability provide some simple applications of Darcy's Law. The constant head permeameter (Figs 5.16, 5.17) follows the obvious route of maintaining a constant flow through a soil sample of known cross-sectional area A and monitoring the pressure at two (or more) points at a known separation – for example, by controlling the total heads at inlet H_1 and outlet H_2 of a soil sample of thickness ℓ. (The filters are provided to contain the soil at top and bottom.) The flow can be measured through the time t taken to collect a known volume V of water. The hydraulic gradient is then

$$i = \frac{H_1 - H_2}{\ell} \tag{5.32}$$

and the permeability is calculated from:

$$k = \frac{V/At}{i} = \frac{V\ell}{At(H_1 - H_2)} \tag{5.33}$$

For soils of lower permeability, a falling head permeameter is often preferred (Fig. 5.18). The inlet head is provided by water in a reasonably narrow tube of cross-sectional area a. The soil sample itself has a considerably larger cross-sectional area A and a thickness ℓ. The head at outlet is maintained constant. Let us take the reference datum at this level so that the total head at outlet is zero. The inlet tube

5.5 Measurement of permeability

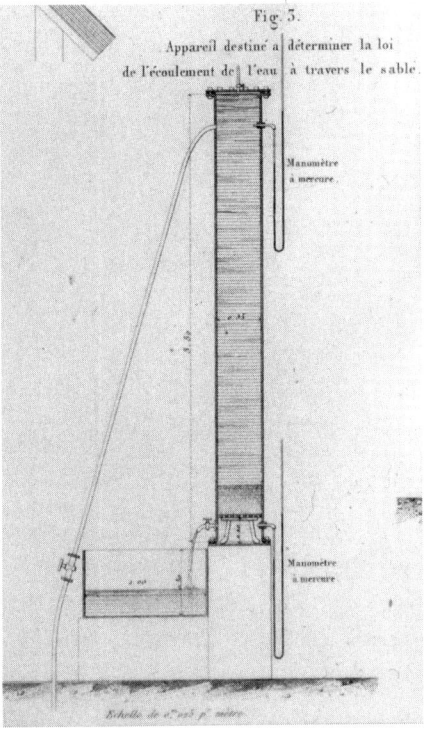

Figure 5.16. Apparatus used by Darcy to determine the permeability of the filters for the water supply of the town of Dijon.[3]

is filled up and the variation of the difference in total head across the sample H is recorded at different times t. The flow rate through the sample at any time is:

$$q = -a \frac{dH}{dt} \tag{5.34}$$

because the flow comes from the falling head of water in the inlet tube of cross sectional area a. The velocity of flow through the sample is:

$$\bar{v} = \frac{q}{A} = -\frac{a}{A}\frac{dH}{dt} \tag{5.35}$$

Applying Darcy's Law, we know that:

$$\bar{v} = ki = k\frac{H}{\ell} \tag{5.36}$$

and rearranging (5.36) and (5.35):

$$\frac{dH}{H} = -\frac{kA}{\ell a} dt \tag{5.37}$$

with solution:

$$\ln \frac{H}{H_i} = -\frac{kA}{\ell a}(t - t_i) \tag{5.38}$$

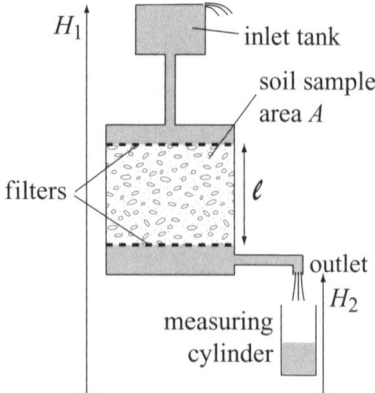

Figure 5.17. Constant head permeameter.

where H_i is the total head in the inlet tube at an initial time t_i (which might be taken as zero). A semilogarithmic plot of inlet head H against time t will have slope $-kA/\ell a$, from which the permeability k can be calculated (Fig. 5.18b). Alternatively, measurements of head H_1 and H_2 at two times t_1 and t_2 can be substituted in (5.38) and permeability calculated directly.

5.6 Permeability of layered soil

It is a nice idealisation to describe ground conditions consisting of thick layers of uniform soil. However, many soils have been formed by deposition over long periods of time with regular or irregular variations in rate of deposition, linked with seasonal or longer climatic cycles (Section 3.5), so that, even though we can still think of conditions being uniform laterally, vertically the properties – especially grain size – will

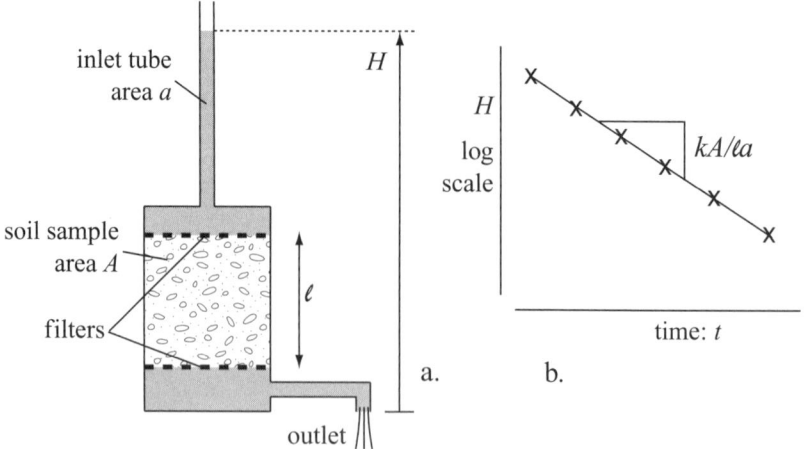

Figure 5.18. (a) Falling head permeameter; (b) variation in head (logarithmic scale) with time (linear scale).

5.6 Permeability of layered soil

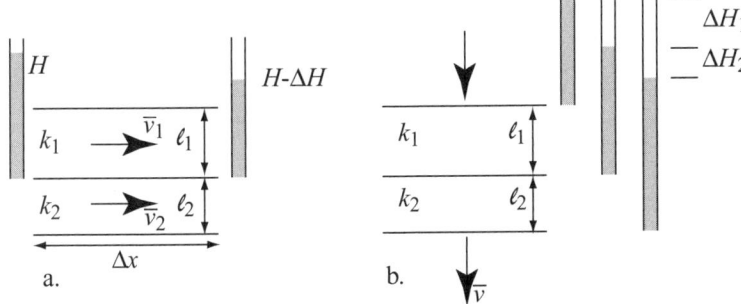

Figure 5.19. Layered soil with (a) horizontal flow; (b) vertical flow.

be somewhat variable. A schematic picture of a two-layered system is shown in Fig. 5.19. The dimensions are not specified: the layer thicknesses might be of the order of millimetres or we might be treating flow at a much larger scale with dimensions of metres. There are two cases to consider: (horizontal) flow parallel to the layering and (vertical) flow orthogonal to the layering.

Parallel flow (Fig. 5.19a) implies (let us suppose) a distribution of total head which varies only in the horizontal direction so that the horizontal hydraulic gradient is the same for all layers. The velocity of flow in layer 1 with thickness ℓ_1 and permeability k_1 is $\bar{v}_1 = k_1 \Delta H / \Delta x$ and the velocity of flow in layer 2 with thickness ℓ_2 and permeability k_2 is $\bar{v}_2 = k_2 \Delta H / \Delta x$. The volume flow rate through the two layers of combined thickness $\ell_1 + \ell_2$ is $q = \bar{v}_1 \ell_1 + \bar{v}_2 \ell_2$ per unit width of the soil (perpendicular to the section in the diagram) and the equivalent horizontal permeability of the layered system is $\bar{k}_h$:

$$\bar{k}_h = \frac{\bar{v}_1 \ell_1 + \bar{v}_2 \ell_2}{\ell_1 + \ell_2} \frac{\Delta x}{\Delta H} = \frac{k_1 \ell_1 + k_2 \ell_2}{\ell_1 + \ell_2} \tag{5.39}$$

If the layers are of the same thickness $\ell_1 = \ell_2$ then the equivalent permeability is the average of the permeabilities of the layers $\bar{k}_h = (k_1 + k_2)/2$. If the ratio of the permeabilities is large – say, $k_1/k_2 = 100$ – then $\bar{k}_h = 0.505 k_1$ and, although the permeability is clearly influenced by the presence of the less permeable layer, the flow is dominated by the more permeable layer. Note that the presence of the less permeable layer has only reduced the average permeability by a factor of about two compared with the ratio of individual permeabilities of 100. And, in fact, our error would not be especially large if we assumed that flow occurred only through the more permeable layers, which make up half the cross section of the soil through which flow is occurring. This would imply an operational, average permeability of $\bar{k}_h = 0.5 k_1$.

Orthogonal flow (Fig. 5.19b) requires that the flow rate through each layer must be the same; the hydraulic gradient must adjust to the individual permeabilities. Thus, the drop in total head ΔH_1 in layer 1 is $\Delta H_1 = \bar{v}\ell_1/k_1$ and the drop in total head ΔH_2 in layer 2 is $\Delta H_2 = \bar{v}\ell_2/k_2$. The average hydraulic gradient through the

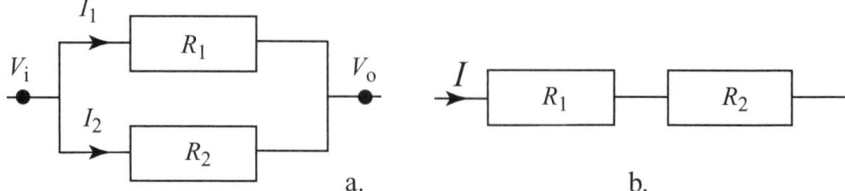

Figure 5.20. (a) Electrical resistors in parallel; (b) electrical resistors in series.

two layers is:

$$\bar{i} = \frac{\Delta H_1 + \Delta H_2}{\ell_1 + \ell_2} = \bar{v}\frac{\ell_1/k_1 + \ell_2/k_2}{\ell_1 + \ell_2} \quad (5.40)$$

and the equivalent vertical permeability of the layered system is $\bar{k}_v$:

$$\bar{k}_v = \frac{\ell_1 + \ell_2}{\ell_1/k_1 + \ell_2/k_2} = \frac{k_1 k_2 (\ell_1 + \ell_2)}{k_2 \ell_1 + k_1 \ell_2} \quad (5.41)$$

If the layers are of the same thickness, then the equivalent permeability is:

$$\bar{k}_v = \frac{2k_1 k_2}{k_1 + k_2} \quad (5.42)$$

If the ratio of the permeabilities is large – say, $k_1/k_2 = 100$ – then $\bar{k}_v = 1.98 k_2$ and the flow rate is primarily controlled by the less permeable layer. The presence of the more permeable layer has only increased the average permeability by a factor of about two compared with the ratio of individual permeabilities of 100. And for this configuration, we could treat the more permeable layer as having infinite permeability $k_1 = \infty$ so that the entire pressure drop occurred only in the low permeability layer. The resulting calculated average permeability would then be $\bar{k}_v = 2k_2$ or, for more general layer thicknesses, $\bar{k}_v = k_2(\ell_1 + \ell_2)/\ell_2$.

The analysis can be extended to a large number of layers of different thicknesses and different permeabilities, but the general conclusions remain: horizontal flow is dominated by the more permeable layers; vertical flow is dominated by the less permeable layers.

There is an obvious analogy with electricity and the resistance of parallel and series networks of resistors. Resistors in parallel (Fig. 5.20a) experience a common drop in voltage while the current is shared between the resistors: the flow of current will be dominated by the lower resistor if the ratio of resistances is large. Resistors in series (Fig. 5.20b) share the same current; the drop in voltage will be dominated by the larger resistor if the ratio of the resistances is large.

5.7 Seepage forces

A gradient of total head is required to drive flow through a narrow tube or through the soil because of the resistance provided by the tube or soil particles trying to impede the flow. Every action has an equal and opposite reaction, so we conclude

5.7 Seepage forces

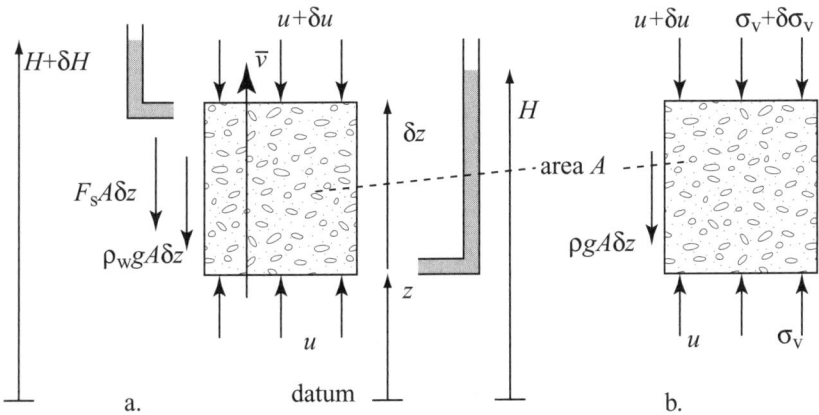

Figure 5.21. Seepage force on soil element.

that the flowing water must in turn be applying a force on the soil through which it is trying to pass. Consider the element of soil in Fig. 5.21 of cross-sectional area A. Water is flowing upwards ($\bar{v} > 0$) and there are algebraic increases in pressure δu and total head δH across the element (although we know that δH must be negative to give this positive direction of flow). The apparent imbalance of forces on the water allows us to find an expression for the seepage force F_s (which we know must be directed upwards), usefully expressed as a *body force* per unit volume of the soil:

$$A\delta u + \rho_w g A \delta z + F_s A \delta z = 0 \tag{5.43}$$

and from the definition of total head δH (5.9):

$$\rho_w g \delta H = \delta u + \rho_w g \delta z \tag{5.44}$$

so that:

$$F_s = -\rho_w g \frac{\delta H}{\delta z} = -\rho_w g i \tag{5.45}$$

where i is the hydraulic gradient – the variation of total head with position. As shown in Fig. 5.21, the hydraulic gradient must in fact be negative to sustain the upward flow and the seepage force as a resistance to the water must be directed downwards.

Alternatively, we can argue that, if there is no flow, then the variation in water pressure across an element exactly matches the variation in elevation head, $\delta u = -\rho_w g \delta z$, and there is no seepage-induced force acting on the soil. So the fact that, as a result of the occurrence of seepage, there is a change in total head then this directly translates into the seepage force exerted by the soil on the water:

$$F_s A \delta z = -\rho_w g A \delta H \tag{5.46}$$

and hence:

$$F_s = -\rho_w g \frac{\delta H}{\delta z} = -\rho_w g i \tag{5.47}$$

The downward resistance on the water becomes an upward force on the soil. In the absence of seepage forces, the vertical total stress across the soil element varies with depth with the bulk density of the soil:

$$\delta\sigma_v = -\rho g \delta z \tag{5.48}$$

However, if the pore pressure gradient is too severe, then the pore pressures may eliminate the normal build-up of total stress with depth. The weight of the element of soil is $\rho g A \delta z$ (Fig. 5.21b). The critical condition comes when the out of balance pore pressure stress $-\delta u = \rho_w g \delta z - \rho_w g \delta H$ (5.44), acting over the area A of the section, exactly balances the weight of the element, which implies:

$$\rho_w g \delta z - \rho_w g \delta H = \rho g \delta z \tag{5.49}$$

which with a little rearrangement becomes an expression for the critical hydraulic gradient, i_{crit}:

$$-\frac{\delta H}{\delta z} = -i_{crit} = \frac{\rho - \rho_w}{\rho_w} = \frac{\rho}{\rho_w} - 1 = \frac{G_s - 1}{1 + e} \tag{5.50}$$

taking the expression for bulk density from Section 3.3. For many soils, the ratio $\rho/\rho_w \sim 2$ and the critical hydraulic gradient $-i_{crit} \sim 1$. The negative sign provides a reminder that it is an upward hydraulic gradient that will cause problems: with our position z measured positive upwards, an upward hydraulic gradient implies that the total head H must be falling with increasing z, or $\delta H/\delta z < 0$.

The same result can be obtained by thinking in terms of effective stresses. If the upward seepage force just balances the buoyant unit weight of the soil, $\gamma' = \gamma - \gamma_w = (\rho - \rho_w)g$, then there is no gradient of effective stress in the soil:

$$-\rho_w \delta H = (\rho - \rho_w)g \delta z \tag{5.51}$$

or:

$$-i_{crit} = -\frac{\delta H}{\delta z} = \frac{\rho}{\rho_w} - 1 \tag{5.52}$$

If there is any tendency of the hydraulic gradient to increase above i_{crit}, then the soil can be lifted up by the flowing water and is said to *pipe*. As the term suggests, the occurrence of a failure in the soil as a result of excessive upward seepage flow will tend to seek out minor weaknesses or inhomogeneities in the soil and form pipes rather than necessarily producing a general instability. But the occurrence of such flow channels, lifting up the soil in the process, will not be conducive to continued satisfactory operation of some hydraulic containment structure (such as a dam or the sheet pile support to an excavation).

On the other hand, downward flow tends to compress the soil and increase the stresses. In fact, downward flow of water through a soil element or structure has been used as a route to the artificial increase in stresses which, to some extent, can eliminate the effects of scale on the stress level in a small model (see Exercise 10 in Section 5.12).

5.8 Radial flow to vertical drain

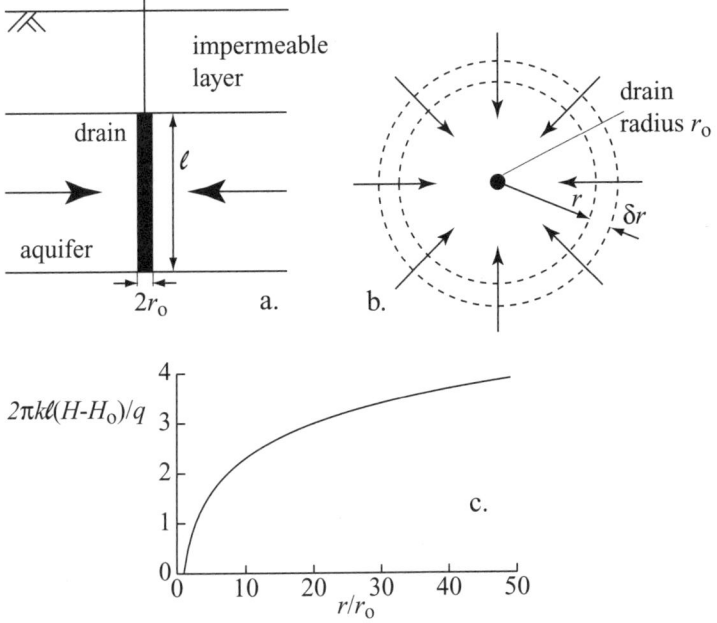

Figure 5.22. Radial flow to vertical drain: (a) section; (b) plan; (c) variation of total head with radius.

5.8 Radial flow to vertical drain

Although we are restricting ourselves to one-dimensional problems, we can be a little adventurous and admit problems which have only a single degree of spatial freedom even though they do not just involve parallel unidirectional flow. There are two problems which are concerned with radial flow with the radius from a drain as the only degree of freedom. Analysis of these problems is instructive anyway because there are many problems in engineering which have radial symmetry and need to be analysed in polar or spherical coordinates (compare the analysis leading to Poiseuille's equation, in Section 5.3).

First, we consider flow to a vertical drain which is being used to lower the total head in a confined aquifer (Fig. 5.22). We have to make the problem tractable so we conveniently specify a drain which extends through the full height of the aquifer, and we make the aquifer confined so that we can reasonably assume that it remains fully saturated even while we pump from the drain.

The governing equation describes the steady flow through concentric annuli (Fig. 5.22b). Conservation of mass requires that the flow rate through each annulus must be the same but the circumference of each annulus depends on its radius, so that the area through which the water is flowing reduces as we get closer to the drain. Hence, the velocity of flow must increase and the hydraulic gradient must increase as we near the drain.

For an aquifer of permeability k and thickness ℓ, the volume flow rate out of the drain is q, or q/ℓ per unit height of drain. The drain has a radius r_o and the total head

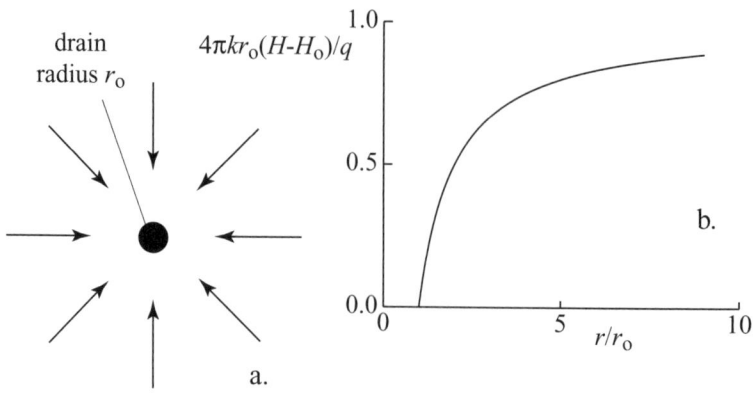

Figure 5.23. Radial flow to point drain.

at the drain is H_o. At radius r, the velocity of flow $\bar{v} = -k\mathrm{d}H/\mathrm{d}r$ (with the negative sign because the water flows down the hydraulic gradient). The volume flow rate through the radius r is $-\bar{v}2\pi r\ell = q$, and, from continuity of flow, the volume flow rate q must be the same at every radius:

$$2\pi r\ell k\frac{\mathrm{d}H}{\mathrm{d}r} = \text{constant} = q \tag{5.53}$$

so that the governing differential equation is:

$$\frac{2\pi k\ell}{q}\mathrm{d}H = \frac{\mathrm{d}r}{r} \tag{5.54}$$

with solution:

$$H - H_o = \frac{q}{2\pi k\ell} \ln \frac{r}{r_o} \tag{5.55}$$

where H_o is the head at the drain. The total head thus varies logarithmically with radius from the drain, as shown in Fig. 5.22c.

5.9 Radial flow to point drain

A similar analysis can be performed for the case of spherical flow to a point drain of radius r_o (Fig. 5.23). Once again, the problem has a single degree of spatial freedom: assuming that we are looking at a volume of soil which is large by comparison with r_o and does not reach any significant layer boundaries or obstacles, everything can vary only with radius r. The same considerations apply to this spherical flow as to the cylindrical flow of Section 5.8: the flow through each spherical layer must be the same and equal to the volumetric outflow rate from the drain q. The cross-sectional area of a layer at radius r is $4\pi r^2$, so our governing equation is:

$$4\pi r^2 k\frac{\mathrm{d}H}{\mathrm{d}r} = q \tag{5.56}$$

5.10 Worked examples: Seepage

so that the governing differential equation is:

$$\frac{4\pi k}{q} dH = \frac{dr}{r^2} \tag{5.57}$$

with solution

$$H - H_o = \frac{q}{4\pi k}\left(\frac{1}{r_o} - \frac{1}{r}\right) = \frac{q}{4\pi k r_o}\left(1 - \frac{r_o}{r}\right) \tag{5.58}$$

and the total head varies with the reciprocal of radius as shown in Fig. 5.23b. Evidently as $r \to \infty$, $H - H_o \to q/4\pi k r_o$.

5.10 Worked examples: Seepage

5.10.1 Example: flow through soil column

Figure 5.24 shows a sample of soil contained in a wide tube or tank. The dimensions shown are evidently somewhat extreme: scepticism needs to be suspended for the sake of this numerical example. The soil is supported on some sort of filter layer so that water can flow in or out of the bottom of the sample without loss of soil particles. The lower part of the tank is connected to an adjacent tube. There is water above the soil and in the adjacent tube. A number of standpipes are shown connected through the wall of the tank containing the soil: two of these (A, B) communicate with the water above the soil sample; two of them (E, F) communicate with the water below the soil sample, and hence with the water level in the adjacent tank; and two of them (C, D) are connected with the soil sample itself, again with suitable filters to prevent loss of soil particles through the connection while ensuring that the water can flow freely. The levels of these standpipes are marked in the figure. Standpipes B and E are located immediately above and immediately below the soil sample, respectively. The measuring points A-F have been chosen at equal vertical separations and Fig. 5.24 has been carefully drawn in such a way that the standpipes are equally spaced horizontally. The line joining the water levels in the several standpipes thus represents exactly the vertical distribution of total head through the soil column. The distribution of total head with position is also shown independently of the standpipes in Fig. 5.24d.

1. First, in Fig. 5.24a, the water levels in the sample tank and the adjacent tube are the same, at a height of 16 m above the reference datum. The water in standpipes A and B has a level equal to the level of the water in the upper reservoir above the soil in the tank at 16 m above the reference datum. The water in standpipes E and F has a level equal to the level of the water in the adjacent tube, thus also to 16 m above the datum. The water pressure or pressure head builds up as we go deeper and deeper below the free water surfaces but, because flow is driven by differences in total head, and not by differences in pressure head, there is no flow through the soil. Since there is no gradient of total head, the total head must be the same also at points C and D, and the

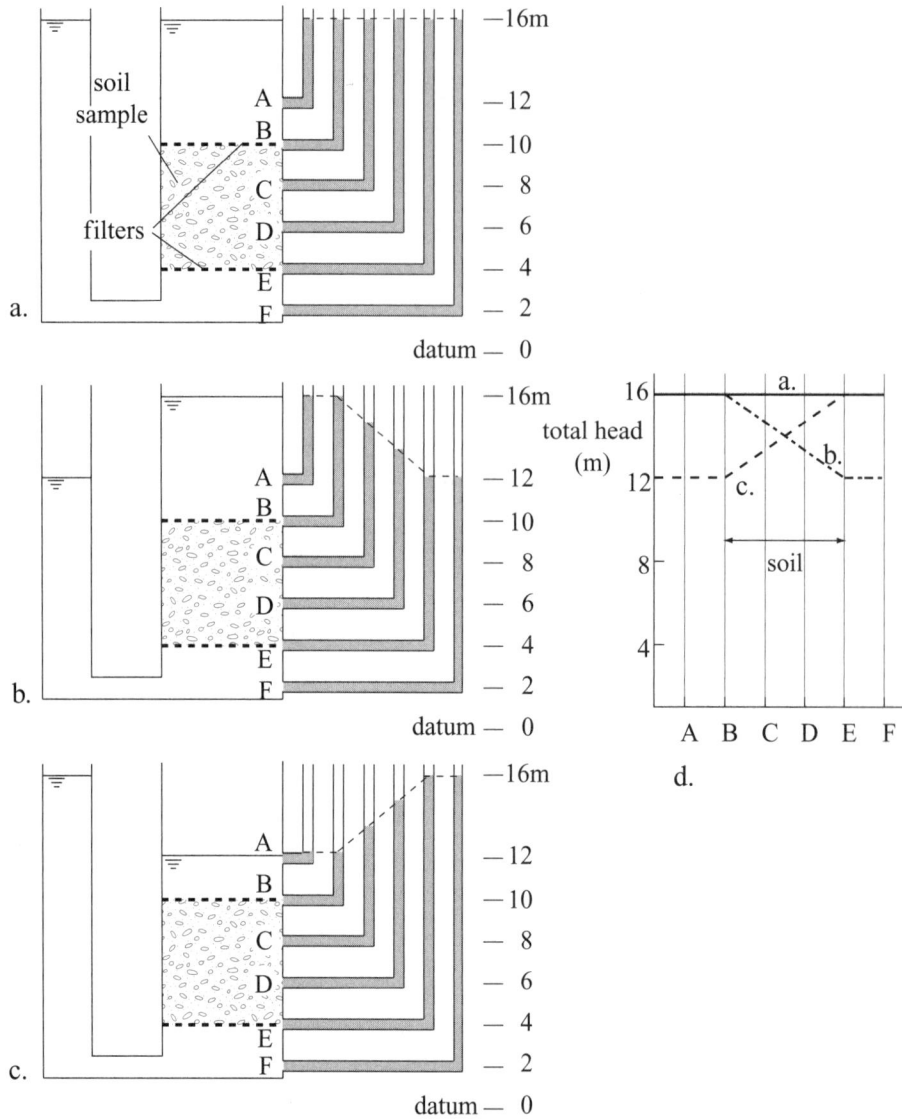

Figure 5.24. Example: flow through soil column.

water in these standpipes rises to the same level. The values of elevation, pressure head and total head for points A-F are shown in Table 5.2. In every case, the pressure head h is simply the difference between the total head H and the elevation z.

2. Second, in Fig. 5.24b, the water level in the adjacent tube is lowered by 4 m to 12 m above the reference datum. We now have a difference in the total heads above and below the soil sample and, as a result, there will be flow from high total head to low total head; in other words, from the large tank through the soil to the adjacent tube. The elevations z of the measurement points A-F do

5.10 Worked examples: Seepage

Table 5.2. *Example: flow through soil column (Fig. 5.24).*

		Fig. 5.24a		Fig. 5.24b		Fig. 5.24c	
	z	H	h	H	h	H	h
Point	m	m	m	m	m	m	m
A	12	16	4	16	4	12	0
B	10	16	6	16	6	12	2
C	8	16	8	14.7	6.7	13.3	5.3
D	6	16	10	13.3	7.3	14.7	8.7
E	4	16	12	12	8	16	12
F	2	16	14	12	10	16	14

not change. The total head at points A and B does not change: the water above the soil has a free surface at 16 m above the datum.

Below the soil, the path through the water to the free surface in the adjacent tube shows that the total head at points E and F is 12 m above the datum. If we assume that the soil is uniform and that the permeability is the same throughout, then, since the cross-sectional area of the soil sample in the tank is constant, the flow velocity must also be constant at all levels. Consequently, we can deduce that the hydraulic gradient must be the same everywhere in the soil sample and thus determine the total heads at C and D by interpolation. Because the measurement points are equally spaced, the drop in total head from B to C is the same as that from C to D and from D to E in Figs 5.24d (chain dotted line) and b. The resulting numerical values are given in Table 5.2. Again, the pressure heads h are determined as the difference between the elevation and the total head at each point. The water pressure head below the soil, at point E, is 8 m and that at the top of the soil sample, at point B, is 6 m but the differences in elevation of points B and E more than compensate for these pressure differences and the flow is downwards, following the gradient of total head (and apparently against the gradient of pore pressure). The hydraulic gradient through the soil is the ratio of the change in total head, $\Delta H = 4$ m, and the thickness of the soil sample over which it occurs, $\Delta z = 6$ m: hence, $i = -\Delta H/\Delta z = -0.667$.

3. Third, in Fig. 5.24c, the water level in the tube is restored to 16 m above the datum and the water level in the tank is lowered to 12 m above datum. We now have a difference in total head above and below the soil sample and, as a result, there will be flow from high total head, below the sample, to low total head, above the sample: in other words, from the adjacent tube into the large tank and upwards through the soil. The elevations z of the measurement points A-F do not change. The total head at points E and F reverts to the original value of 16 m above datum: follow the path through the water to the nearest free surface in the adjacent tank. Above the soil, the measurement points A and B are directly aware of the water level in the tank and the total head at points A and B is 12 m above the datum. We can again deduce that the hydraulic gradient must be the same everywhere in the soil sample, because the flow is

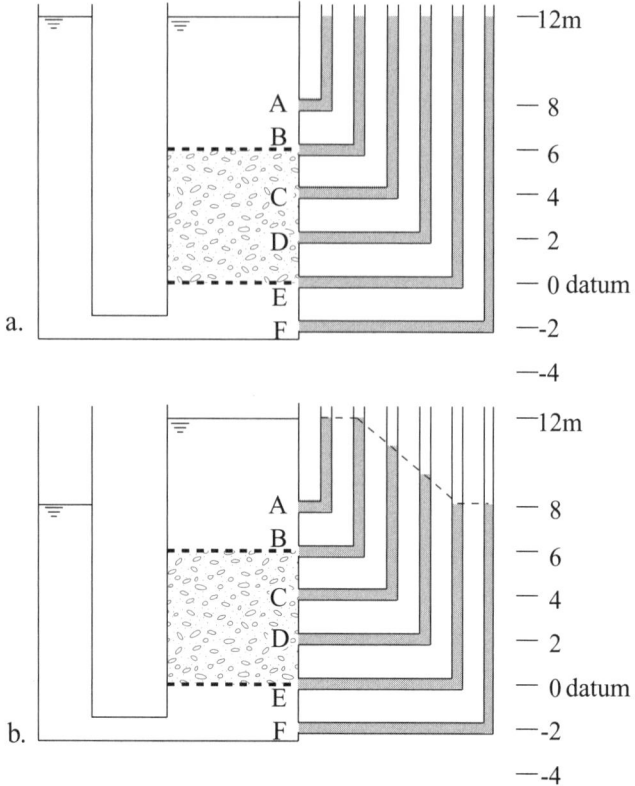

Figure 5.25. Example: flow through soil column with changed reference datum (compare Fig. 5.24).

steady and the area of the cross section is constant, and thus determine the total heads at C and D by interpolation. Because the measurement points are equally spaced, the rise in head from B to C is the same as that from C to D and from D to E in Figs 5.24d (dashed line) and c. The resulting numerical values are given in Table 5.2. Again, the pressure heads h are determined as the difference between the elevation and the total head at each point. The hydraulic gradient is calculated in the same way: $\Delta H = -4$ m, $\Delta z = 6$ m, $i = 0.667$ and the flow is upwards.

5.10.2 Example: effect of changing reference datum

It is important to recall that elevation z and total head H only have meaning with reference to some datum. The choice of location of this datum is often arbitrary and may be chosen for convenience to ensure, for example, that all heads are positive. It is necessary only that the reference datum should be the same for all points and the same for calculation of both elevation and total head. Figure 5.25 repeats the configurations of Figs 5.24a and b but with the datum chosen at the base of the soil

5.10 Worked examples: Seepage

Table 5.3. *Example: flow through soil column (Fig. 5.25).*

		Fig. 5.25a		Fig. 5.25b	
Point	z m	H m	h m	H m	h m
A	8	12	4	12	4
B	6	12	6	12	6
C	4	12	8	10.7	6.7
D	2	12	10	9.3	7.3
E	0	12	12	8	8
F	−2	12	14	8	10

sample in the tank instead of 2 m below the base of the tank. The corresponding values of elevation, total head and pressure head are shown in Table 5.3.

The consideration of this problem need not detain us long. We have lifted the datum by 4 m; consequently, all elevations and all total heads are 4 m lower than in Table 5.2. Differences in heads referred to the datum are unchanged so the hydraulic gradient across the soil sample has not changed: the difference between the total heads at points B and E is the same in Tables 5.2 and 5.3. Pressure head is calculated as a difference between quantities referred to the datum and the values do not change with change of datum. Change in the choice of datum does not affect the conclusion that flow occurs downwards from top to bottom of the sample.

5.10.3 Example: pumping from aquifer

A schematic diagram of the ground conditions at a particular site is shown in Fig. 5.26a. A layer of sand overlies a layer of silty clay, which in turn overlies a sand aquifer. An *aquifer* is a water-bearing layer which usually has some spatial connectivity to a natural or artificial source of water which controls the water pressure in the layer. The water table is naturally at a depth of 2 m below the ground surface, and this represents also the total head for the aquifer assuming that we do not interfere with the groundwater conditions. We consider two configurations.

1. First, the water is static and not flowing. By definition, the total head H will be constant with depth, with height 24 m above the datum as shown by the solid line in Fig. 5.26b. The elevation head z increases linearly from zero at the datum as shown by the dashed line in Fig. 5.26b: it matches the total head at the level of the water table. The pressure head $h = u/\rho_w g$ is the difference between the total and elevation heads and varies linearly with depth, as shown by the chain dotted line in Fig. 5.26b. The pressure head is zero at the water table and is 24 m at the level of the datum.
2. Second, we consider the case when there is vigorous pumping from the aquifer to provide water for a small town such that the total head in the aquifer falls to 18 m above the datum (Fig. 5.26c). The water table in the upper sand layer remains unchanged – it is fed from some near surface water sources such as

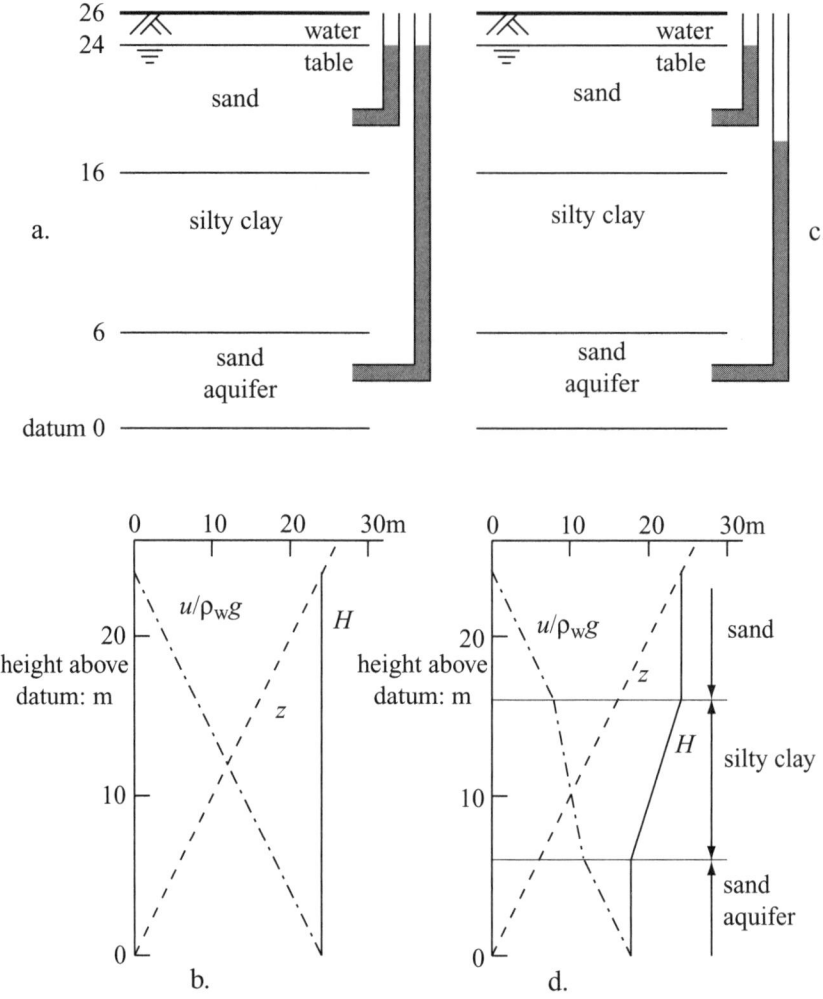

Figure 5.26. (a, b) Ground conditions and heads (heights above datum in metres) without flow; (c, d) ground conditions and heads with pumping from aquifer.

a river or lake. The silty clay is able to sustain the difference in total head and, after a while, a steady seepage regime is established (there is a transient phase during which the new flow conditions settle down – this is the subject of Chapter 7).

The variation of total head with depth now has three sections, as shown in Fig. 5.26d. In the aquifer, the lower sand layer, the total head is constant at 18 m above the datum. In the upper sand layer, the total head is constant at 24 m above the datum (as in the no-flow condition). Through the silty clay the flow rate must be constant (from conservation of mass), so the hydraulic gradient must be uniform and the total head must vary linearly with depth. There is a 6 m drop in total head over a thickness of 10 m of silty clay, so that the constant hydraulic gradient through the silty clay is 0.6. The tripartite variation of H is

5.10 Worked examples: Seepage

shown with the solid line in Fig. 5.26d. The elevation head z has not changed, so the variation of z with depth is the same as it was in the no-flow condition: the dashed line in Fig. 5.26d is the same as the dashed line in Fig. 5.26b. The pressure head $h = u/\rho_w g$ is, as usual, the difference between total head H and elevation head z, and it is shown by the chain dotted line in Fig. 5.26d. The pressure head at the top of the silty clay is 8 m, as before. At the bottom of the silty clay it is 12 m (compared with 18 m before pumping): the total head has been reduced to 18 m but the elevation head is still 6 m. The pressure head varies linearly through the silty clay.

If we know the permeability of the silty clay – let us say that it is 10^{-8} m/s – then we can calculate the flow rate through this layer. The hydraulic gradient is $i = \Delta H/\Delta z = 6/10 = 0.6$ and the flow rate is therefore 0.6×10^{-8} m^3/s per m^2. For a site with area 1 hectare (100×100 m^2) the flow rate is therefore 0.06 litres/s or 5184 litres/day.

5.10.4 Example: flow into excavation

Since we are restricting ourselves to simple one-dimensional flow configurations, we are limited in the complexity of the prototypes that we can model, and every case that we consider will really be just a modified version of the problem considered in Section 5.10.3 and Fig. 5.26 – we can use "corroborative detail ... to give artistic verisimilitude to an otherwise bald and unconvincing narrative" but the basic one-dimensional analysis techniques remain applicable.

Thus, Fig. 5.27 shows an excavation through fill and silty sand near a river to form some sort of temporary space within which construction work can, hopefully, proceed in the dry. The excavation is protected with a "cofferdam" formed of steel sheet piles (Fig. 1.11). We are interested to understand the water pressure conditions in the silty sand and in the underlying gravel aquifer as the water levels in the excavation and in the nearby river change. We consider four monitoring points: point P at a level -6 m relative to the datum, and points X, Y, Z at the top, middle and bottom of the silty sand, at elevations 0 m, -2 m and -4 m relative to the datum. The diagram is intended to indicate that the river is in hydraulic communication with the aquifer so that the water level in a standpipe placed anywhere in the gravel (such as P) will always match the level of the water in the river.

1. First (Fig. 5.27b), we consider the case where the water level in the excavation and the water level in the river are both at +1 m relative to the datum (the pumps have not been switched on to keep the excavation fully dry). The total head in the aquifer is at +1 m so the total head at points P and Z must be +1 m: Z is just at the interface between the gravel and the silty sand. At the top of the silty sand, the nearest free water surface is provided by the water in the excavation and the total head is thus also +1 m: imagine inserting a standpipe at X and a long flexible standpipe from P up through the river. There is therefore no variation of total head across the silty sand, the total head is everywhere

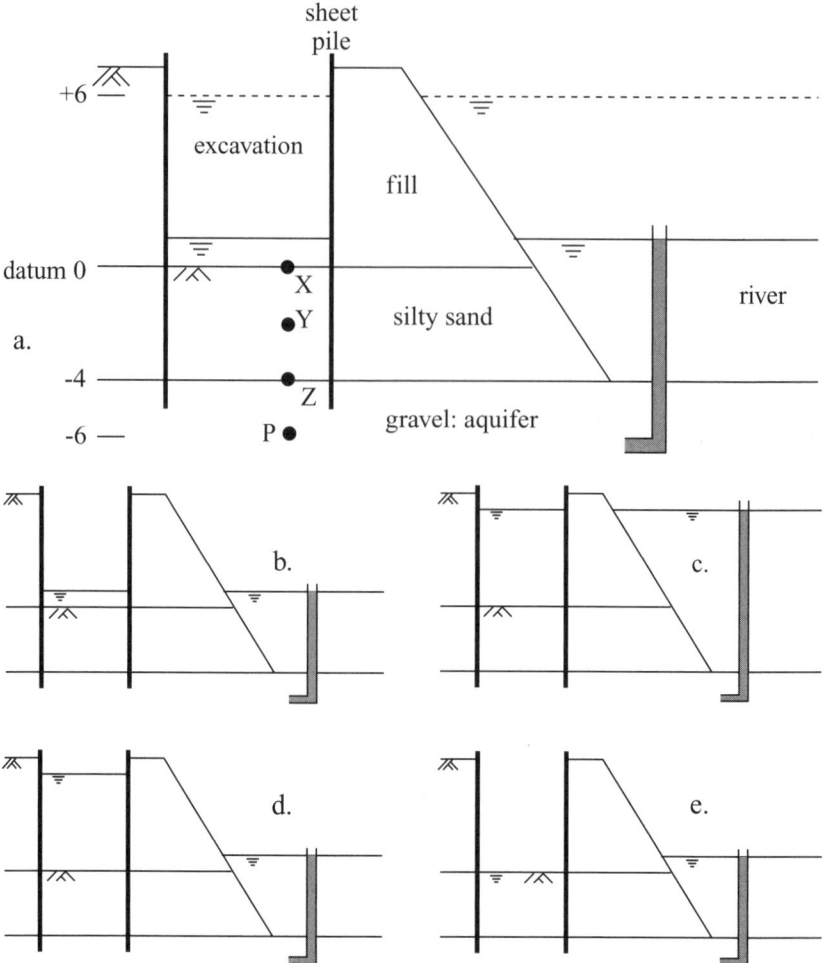

Figure 5.27. Excavation near river (heights above datum in metres).

+1 m and there is no flow. The values of elevation, total head and pressure head are listed in Table 5.4.

2. Second (Fig. 5.27c), we consider the case where the water level in the excavation and the water level in the river are both at +6 m relative to the datum: there has been a failure of the pumps keeping the excavation dry during a period of exceptionally heavy rainfall and the water level has simply followed the level of the river, which is in flood. The total head in the aquifer is +6 m: this controls the total head at points P and Z. The total head in the excavation is +6 m: this controls the total head at point X. There is no variation of total head across the silty sand: the total head is everywhere +6 m and there is no flow. The corresponding values of total head and pressure head are listed in Table 5.4.

3. Third (Fig. 5.27d), we consider the case where the river level has fallen back to its normal level of +1 m but the excavation remains flooded to the level +6 m. The total head at point X is now +6 m: the water level in the excavation is the

5.11 Summary

Table 5.4. *Example: excavation near river (Fig. 5.27).*

		Fig. 5.27b		Fig. 5.27c		Fig. 5.27d		Fig. 5.27e	
	z	H	h	H	h	H	h	H	h
Point	m	m	m	m	m	m	m	m	m
X	0	+1	1	+6	6	+6	6	0	0
Y	−2	+1	3	+6	8	+3.5	5.5	+0.5	2.5
Z	−4	+1	5	+6	10	+1	5	+1	5
P	−6	+1	7	+6	12	+1	7	+1	7
flow?		no		no		down		up	

nearest free water surface. The total head at points P and Z is back to +1 m, so that there is a drop of 5 m in the total head across the silty sand. The gradient of total head will lead to downward flow taking place through the silty sand. Because the flow velocity must be uniform, the gradient of total head through the silty sand must be constant and the value of the total head at Y can be determined as +3.5 m by linear interpolation between the values for points X and Z. The pressure heads can be calculated from the total heads and elevations as usual; they are listed in Table 5.4. The hydraulic gradient downwards through the silty sand is $i = -\Delta H/\Delta z = 5/4 = 1.25$.

4. Fourth (Fig. 5.27e), we consider the case where the river level is at its usual level of +1 m and the excavation is being pumped dry, so that the water level is coincident with the base of the excavation at level 0 m, coinciding with the datum. Construction works can at last proceed in the dry as planned. The total head at point X is now 0 m. The total head at points P and Z is +1 m. There is thus an upward hydraulic gradient across the silty sand leading to upward flow. The total head at point Y is found by linear interpolation between the values at X and Z: it is +0.5 m. The pressure heads can be calculated from the total heads and elevations as usual; they are listed in Table 5.4. The hydraulic gradient upwards through the silty sand is $i = -\Delta H/\Delta z = -1/4 = -0.25$.

Given a value for the permeability of the silty sand, say 10^{-5} m/s, and a width for the excavation, say 6 m, we can calculate the flow rate down from the excavation in the third case and up into the excavation in the fourth case. With the river at +1 m and the total head in the excavation at +6 m, the hydraulic gradient is 1.25 and the flow rate is $1.25 \times 10^{-5} \times 6 = 7.5 \times 10^{-5}$ m^3/s per metre length of excavation. With the river at +1 m and the total head in the excavation at 0 m, the hydraulic gradient is 0.25 and the flow rate is $0.25 \times 10^{-5} \times 6 = 1.5 \times 10^{-5}$ m^3/s per metre length of excavation.

5.11 Summary

Here is a concise list of the key messages from this chapter, which are also encapsulated in the mind map (Fig. 5.28).

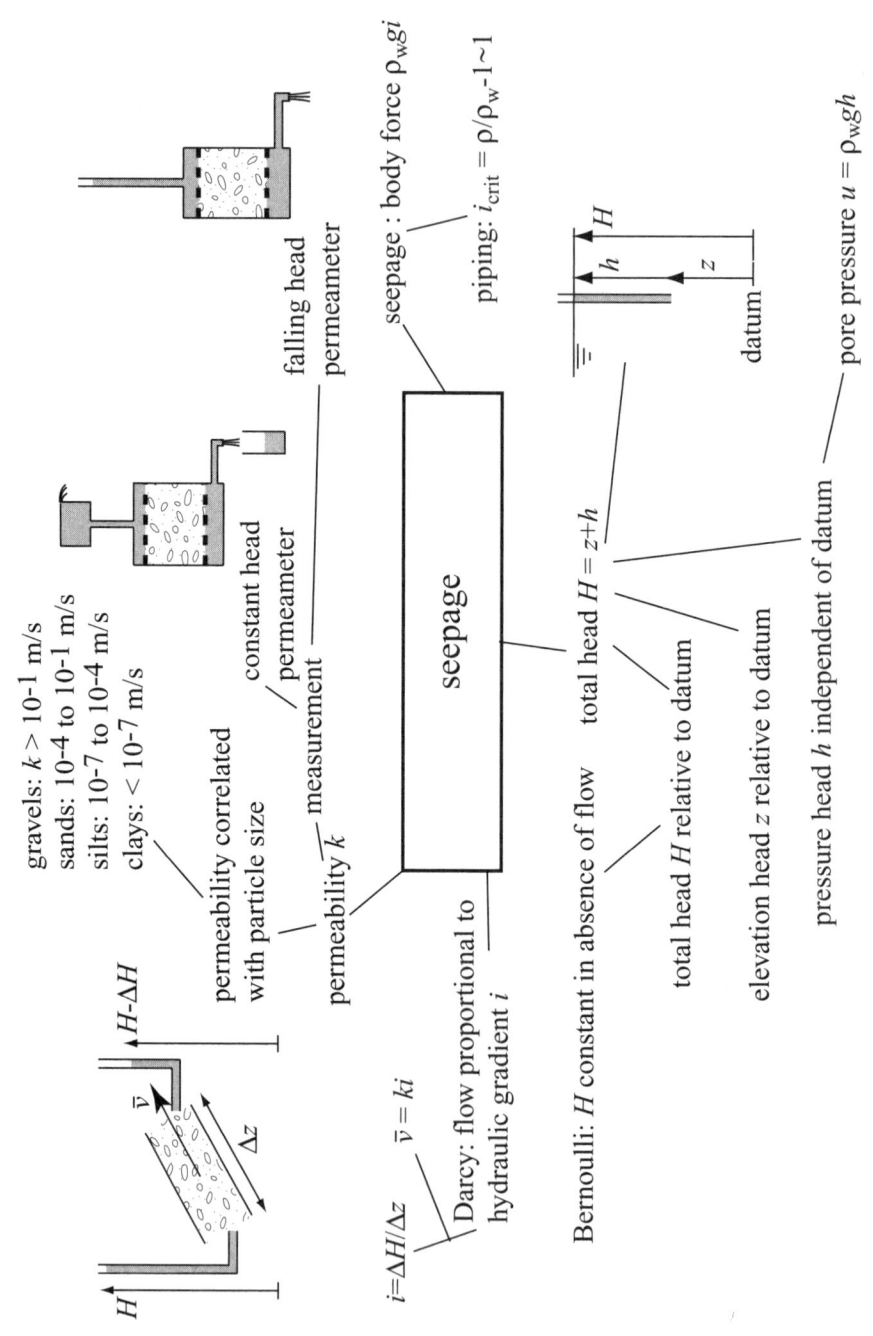

Figure 5.28. Mind map: seepage.

5.12 Exercises: Seepage

1. Total head H is defined as the sum of elevation head z and pressure head $h = u/\rho_w g$. Total head and elevation head must be referred to a datum; pressure head is independent of datum.
2. Flow is driven by gradients of total head H: hydraulic gradient $i = dH/dz$.
3. Darcy's Law relates velocity of flow with hydraulic gradient, introducing a constant of proportionality, the permeability of the soil k.
4. Permeability is closely correlated with particle size: the permeability of gravels may be 8 to 10 orders of magnitude greater than the permeability of clays. Hence, gravels are used for drains and clays are used for barriers to flow.
5. Permeability can be measured using constant head or falling head configurations.
6. Flow of water through soils generates a seepage force, a body force $\rho_w g i$ per unit volume; the critical upward hydraulic gradient for piping uplift failure of the soil – in which the upward seepage force balances the weight of the soil – is of the order of unity.
7. Flow through layered soil parallel to the layers is dominated by the most permeable layers; flow orthogonal to the layers is dominated by the least permeable layers.
8. Analysis of radial flow under conditions of cylindrical or spherical symmetry produces simple expressions for the nonlinear variation of total head with radius.

5.12 Exercises: Seepage

1. A vertical tube containing fine sand with permeability 0.05 m/s is inserted into a tank of water (Fig. 5.29). Initially, the water levels are the same in the tube and the tank. The column of sand is 0.6 m high and it is held with the base of the sand (Z) 0.2 m above the base of the tank and the top of the sand (X) 0.2 m below the water surface. Plot the variation of total head, pressure head and elevation head with height. Find the pressures at points X, Y, Z at the top, centre, and base of the sand.

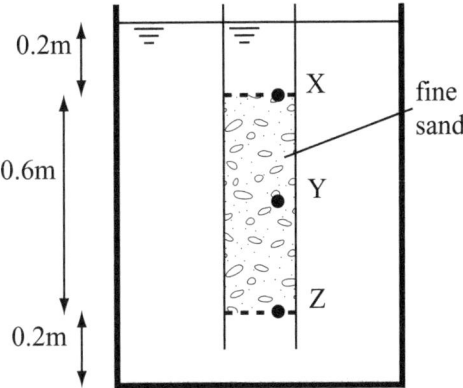

Figure 5.29. Diagram for Question 1.

The water level is then lowered quickly in the tank by 0.8 m. The water level in the tube above the soil is maintained at its previous level by filling from the main water supply. Repeat the plot of variation of heads and again find the pressures at points X, Y, Z. Does the water move through the sand? If so, in which direction and at what speed?

2. At a certain site, 5 m of sand overlie 6 m of clay (with permeability 10^{-7} m/s), which overlie a sand aquifer 3 m thick. The datum is at the base of the aquifer. The water table is 1 m below the ground surface. Plot the variations of total head, elevation head and pressure head (a) for the situation described in which there is no flow; and (b) for an alternative situation when pumping from the lower sand aquifer reduces the total head in that layer to $+10$ m while the water table remains unchanged in the upper sand layer. Calculate the long-term seepage velocity through the clay.

3. At a certain location, the ground conditions consist of 10 m sand, over 10 m clay, over 10 m sand and gravel (which constitutes a lower aquifer), over bedrock. The water table in the upper sand layer is 5 m below ground level. Plot the distribution of total head and pressure head against depth, and hence find the distribution of pore water pressure: (a) if ground water conditions are hydrostatic; (b) if the piezometric level in the lower aquifer is reduced by 5 m by pumping, leaving the level in the upper layer unaltered. (Hint: Within each of the permeable layers (sand and gravel aquifers) conditions are hydrostatic, with constant total head. Can you explain why?)

For each case, find the water pressure at the centre of the clay layer. For case (b) calculate the flow rate: the clay has permeability 2×10^{-7} m/s.

4. Figure 5.30 shows a cross-section through a river bank where there is an embankment, 3 m high, to reduce the risk of flooding. The gravel layer is hydraulically connected to the river so that the piezometer P always indicates the same level as the river.
   - Normal river level is 1 m below ground level, as shown.
   - When the river level is at or below ground level, the water level in the silty sand matches the river level.

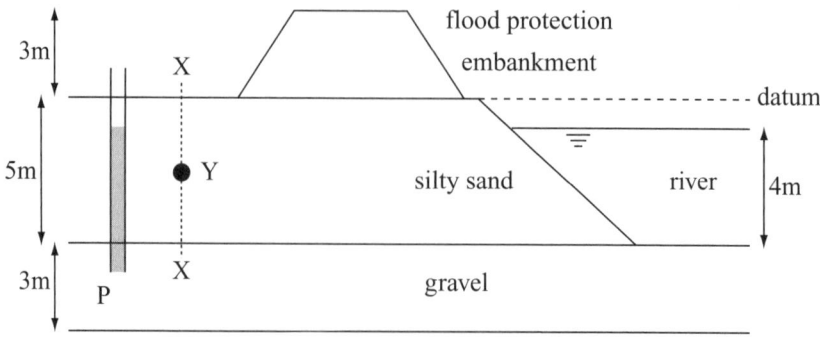

Figure 5.30. Diagram for Question 4.

5.12 Exercises: Seepage

- The land behind the flood embankment is usually drained to no higher than ground level unless catastrophic overtopping of the embankment occurs.

Plot distributions of total head and water pressure with depth at the section XX and determine the water pressure at point Y at the centre of the silty sand layer:

a. When the river level is normal at −1 m (and the water level in the silty sand is also at −1 m);
b. When the river level is at 0 m (and the water level in the silty sand is also at 0 m);
c. When the river level is just below +3 m (and drainage keeps the water level in the silty sand at 0 m);
d. When the river level is just above +3 m and the drains are overloaded, so that the land beyond the embankment is flooded to a depth of 3 m;
e. When the river level has fallen back to -1 m, leaving the drains overloaded and the land beyond the embankment flooded to +3 m.

In each case, indicate whether water is flowing through the silty sand and, if so, in which direction and at what rate. The permeability of the silty sand is 10^{-4} m/s. Assume that the seepage through the silty sand is vertical and that a condition of steady flow has been attained.

5. A layer of silty clay 4 m thick with permeability $k = 10^{-7}$ m/s is sandwiched between two layers of sand, each 4 m thick, with permeability $k = 10^{-2}$ m/s. The water table in the upper layer of sand is at the ground surface. The lower layer of sand has an artesian pressure 3 m above ground level. Calculate the pressure head and pore pressure at the top and bottom of the silty clay and calculate the flow rate through the silty clay.

6. A layer of sand 4 m thick with permeability $k = 10^{-2}$ m/s is sandwiched between two layers of silty clay each 4 m thick with permeability $k = 10^{-7}$ m/s. The lower layer of silty clay is underlain by a gravel aquifer under artesian pressure with total head 3 m above ground level. The total head at the top of the upper layer of silty clay is at ground level. What are the total head (relative to ground level) and pressure head at the top and bottom of the layer of sand? What is the flow rate through the silty clay? What is the drop in total head across the sand layer?

7. Seasonal deposition has produced a varved deposit which consists of alternate layers of silt of thickness 1 mm and of fine sand of thickness 2 mm, with permeabilities 5×10^{-5} m/s and 2×10^{-3} m/s, respectively. Calculate the permeabilities of the varved soil to horizontal and vertical flow.

8. A layer of clay 4 m thick (permeability $k = 10^{-8}$ m/s) lies between two layers of sand each 3 m thick, the top of the upper layer of sand being at ground level. The water table is 1m below the ground level but the lower layer of sand is under artesian pressure, the total head being 3 m above ground level. The saturated density of the clay is 2 Mg/m^3 and that of the sand is 1.9 Mg/m^3; above the water table, the sand has density 1.65 Mg/m^3. Calculate the total head (relative to ground level), pressure head, pore pressure, total stress and effective stress

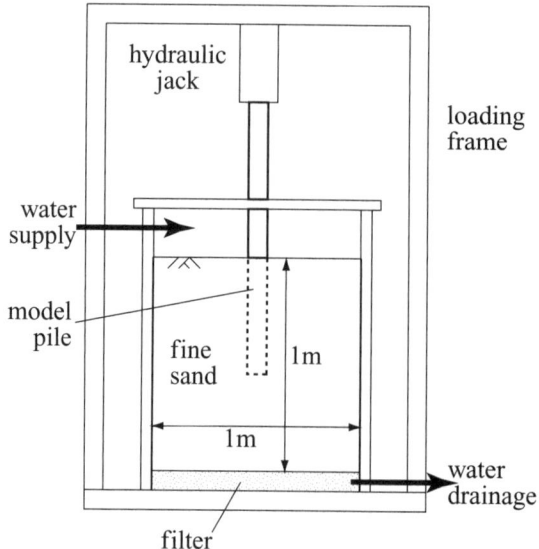

Figure 5.31. Diagram for Question 10.

at the top and bottom of the clay layer. Calculate the flow rate through the clay layer.

9. In a falling head permeability test, the initial head of 1.00 m dropped to 0.25 m in 4 h, the diameter of the standpipe inlet being 5 mm. The length of the soil specimen was 200 mm and its diameter 100 mm. Calculate the permeability of the soil.

10. A small physical model is prepared for a laboratory study of the load transfer mechanism for driven piles in fine sand. Laboratory space dictates that the model can be no more than 1 m diameter and 1 m high, whereas the prototype layer of soil being modelled is about 20 m deep with the water table at the ground surface. There is concern that the low stress level in the laboratory may distort the experimental observations, and it is proposed to apply a downward hydraulic gradient to the model in order to establish prototype stress levels at the base of the soil (Fig. 5.31).[6]

The sand has void ratio 0.55 and the specific gravity of the soil mineral is 2.65. The permeability of the sand is $k = 10^{-4}$ m/s.

What hydraulic gradient is required? What flow rate is required to maintain this hydraulic gradient?

---

[6] Zélikson, A. (1969) Geotechnical modelling using the hydraulic gradient similarity method. *Géotechnique* **19** 4, 495–508.

6 Change in stress

6.1 Introduction

In Chapter 4 we introduced the concept of the stiffness of soils under one-dimensional loading and were able to calculate the change in vertical strain, and hence the change in thickness of a soil layer that might occur as a result of a change in effective stress. In Chapter 5, we encountered the concept of permeability of soils and noted in particular the huge range of values of permeability for soils, broadly, of different particle sizes (but also influenced by the mineralogy and shape of the particles). The permeability of clays is many orders of magnitude lower than the permeability of sands and gravels. Change in effective stress implies change in vertical dimension of the soil layers, which implies the squeezing out or the sucking in of water (assuming that the soil is saturated). In a soil of very low permeability this cannot happen rapidly, and, in this chapter, we will make deductions about the short-term and long-term conditions that must apply. The analysis of the process that spans between the short term and the long term is called *consolidation* and is the subject of Chapter 7.

6.2 Stress change and soil permeability

Figure 6.1 provides an analogy for the behaviour of a soil with low permeability when it is subjected to a change in external stresses. The spring represents the soil, and the loads taken by the spring represent the effective stress carried by the soil particles. The spring is contained in a cylinder (like an oedometer) of cross sectional area A, which is full of water (the pore fluid in the soil) and capped by a rather tightly fitting piston. When a sudden change in external load P is applied to the piston, the spring would like to compress. However, for the spring to compress water needs to escape past the piston, and this cannot occur rapidly. Equilibrium tells us that the load P must be carried somehow: it cannot be carried by the spring unless the spring compresses, so it must be carried by the water. The water is comparatively incompressible (which means that its stiffness for one-dimensional confined deformation is pretty high) so if we call the pressure in the water u then, immediately

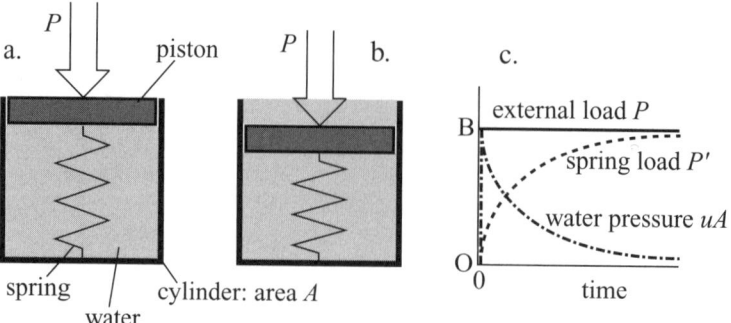

Figure 6.1. (a, b) Spring and tightly fitting piston; (c) gradual transfer of applied load from pore pressure to spring load.

after the load P is applied, the force provided by the water pressure acting over the cross-section A of the cylinder is:

$$uA = P \tag{6.1}$$

as indicated by the chain dotted line OB at time $t = 0$ in Fig. 6.1c.

Now the very narrow gaps around the piston, which are analogous to the fine and tortuous pores in our low permeability soil, are also broadly equivalent to the narrow tubes to which Poiseuille's equation refers (Section 5.3). With a pressure gradient across the thin annular gap, flow will occur at a rate proportional to the pressure difference between the water inside the cylinder and the free water sitting on top of the piston. As flow occurs, so the spring will be able to compress and, as time goes on, the external load P will be completely transferred to the spring and the pressure in the water will ultimately be zero again. If we call the force in the spring at any time P', then we can propose that at all times during this process:

$$P = uA + P' \tag{6.2}$$

with the limiting conditions that, for $t = 0$, $P' = 0, u = P/A$ and, as $t \to \infty$, $P' \to P$ (dashed line) and $u \to 0$ (chain dotted line) (Fig. 6.1c). We will call this pore pressure u, which is tending to disappear with time, the *excess* pore pressure. In this chapter, we will concern ourselves only with these two limiting conditions: $t = 0$ and $t \to \infty$.

Exactly the same sort of thought process can be used to describe what happens when the external total vertical stress σ_z acting on the loading platen of an oedometer containing a sample of clay (Fig. 6.2) is changed. The deformation of the soil requires change in the volumetric packing of the soil particles. The voids around the soil particles are filled with water, so that change in the packing can only occur if the water is able to flow through the voids. The permeability of this clay sample is low, so there is no possibility of immediate change in volumetric packing. Equilibrium tells us that at all times the total vertical stress can only be supported by a combination of effective stress σ'_z between the soil particles – the "soil spring" – and pressure

6.2 Stress change and soil permeability

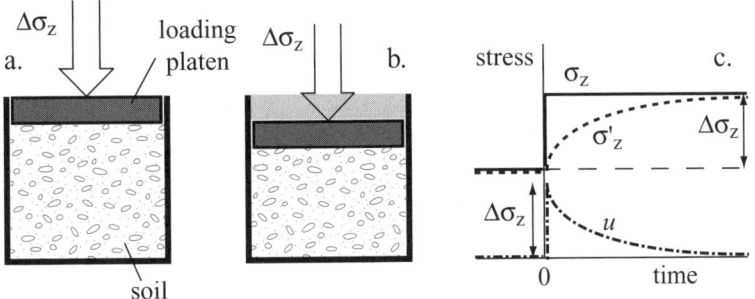

Figure 6.2. (a, b) Clay sample in oedometer; (c) gradual transfer of change in total stress $\Delta\sigma_z$ from pore pressure to effective stress.

u in the water surrounding the soil particles:

$$\sigma_z = \sigma'_z + u \tag{6.3}$$

and this subdivision applies equally for changes in each of the stress components:

$$\Delta\sigma_z = \Delta\sigma'_z + \Delta u \tag{6.4}$$

At the moment that the total stress is changed there is no possibility of the effective stress changing, and consequently the change in external total vertical stress $\Delta\sigma_z$ must be converted into a change in pore water pressure Δu: $\Delta u = \Delta\sigma_z$ (Fig. 6.2c).

Just as for the spring in the cylinder (Fig. 6.1), with time the water is able to flow and the soil spring is able to compress and the effective stress is able to take up the change in external stress $\Delta\sigma'_z = \Delta\sigma_z$. Our limiting conditions become, as before: $t = 0$, $\Delta\sigma'_z = 0$, $u = \Delta\sigma_z$ and, as $t \to \infty$, $\Delta\sigma'_z \to \Delta\sigma_z$ and $u \to 0$ (Fig. 6.2c).

The terms *undrained* and *drained* are used to describe these two extremes of response. *Undrained* signifies the condition immediately after the change in external loading at time $t = 0$ before the water has had time to move or drain from the voids in the soil. *Drained* signifies the long-term establishment of pore water equilibrium – which may be an equilibrium of steady flow, but a flow regime that does not change with time – and the transfer of total stress from initially non-equilibrated pore pressure to effective stress. The classification of events as undrained or drained obviously depends on the soil type – primarily, its permeability – and the nature of the loading. If we propose that sands and gravels, having high permeability, will almost invariably respond in a drained manner because the pore water can flow easily, we must accept that even for such soils there may be occasions when the rate at which the external load is changing is so high – as a result of an explosion, for example, or an earthquake – or the distance over which the water has to flow to reach freedom is so great, that the pore pressure changes from its equilibrium value and the response, even for a short period, is that of an undrained soil.

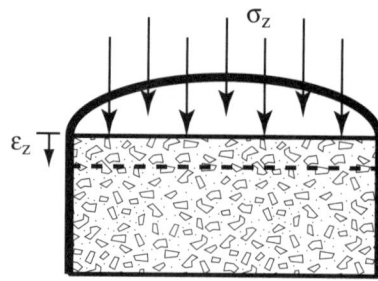

Figure 6.3. Soil sample in oedometer (Section 6.3.1). Compare Figs 4.5, 4.6.

6.3 Worked examples

6.3.1 Example 1

Let us first look at a simple example: a sample of clay in an oedometer (Fig. 6.3). Let us suppose that the current externally applied vertical total stress $\sigma_z = 200$ kPa is in equilibrium with the effective stress σ'_z carried by the soil particles and that the pore pressure is zero, $u = 0$, so that $\sigma'_z = \sigma_z - u = 200$ kPa.

1. At time $t = 0$, we increase the external stress by $\Delta\sigma_z = 100$ kPa to $\sigma_z = 300$ kPa. The clay is not able to deform rapidly so the effective stress is not able to change and the increase in total stress is transferred to the pore pressure: $\Delta u = \Delta\sigma_z = 100$ kPa and $\Delta\sigma'_z = 0$.

   In the long term, the pore pressure is allowed to dissipate so that, once again, $u = 0$. The externally imposed vertical total stress has not changed, so the vertical effective stress is calculated as the difference between the total stress and the zero pore pressure, $\sigma'_z = 300$ kPa, and the eventual change in effective stress matches the original change in total stress at time $t = 0$. Thus, after a long time ($t \to \infty$), $\Delta\sigma'_z = 100$ kPa.

2. Exactly the same procedure can be followed when the total stress is lowered rather than increased. The clay would now like to increase in volume as the stress decreases but the low permeability prevents this happening rapidly: the pore pressure has to take a negative value to hold back the loading cap of the oedometer. Suppose the total stress is lowered by $\Delta\sigma_z = -50$ kPa so that $\sigma_z = 250$ kPa. The effective stress is unchanged at $\sigma'_z = 300$ kPa. The pore pressure is as ever given by the difference between total and effective stresses: $u = \sigma_z - \sigma'_z = 250 - 300 = -50$ kPa.[1]

   Once again, though, if we leave the oedometer long enough, the pore pressure will be able to re-establish its equilibrium value $u = 0$ and the long-term change in effective stress will match the reduction of total stress: $\Delta\sigma'_z = \Delta\sigma_z = -50$ kPa, so that $\sigma'_z = \sigma_z = 250$ kPa.

---

[1] The negative sign for the pore pressure should not worry us – though if the negative pore pressure approaches the value of atmospheric pressure, around 100 kPa, then the absolute pressure in the pore fluid becomes close to zero and there is a danger that cavitation of the pore water may occur. Air bubbles would then come out of solution and the assumption that the pore fluid is incompressible would no longer be tenable.

6.3 Worked examples

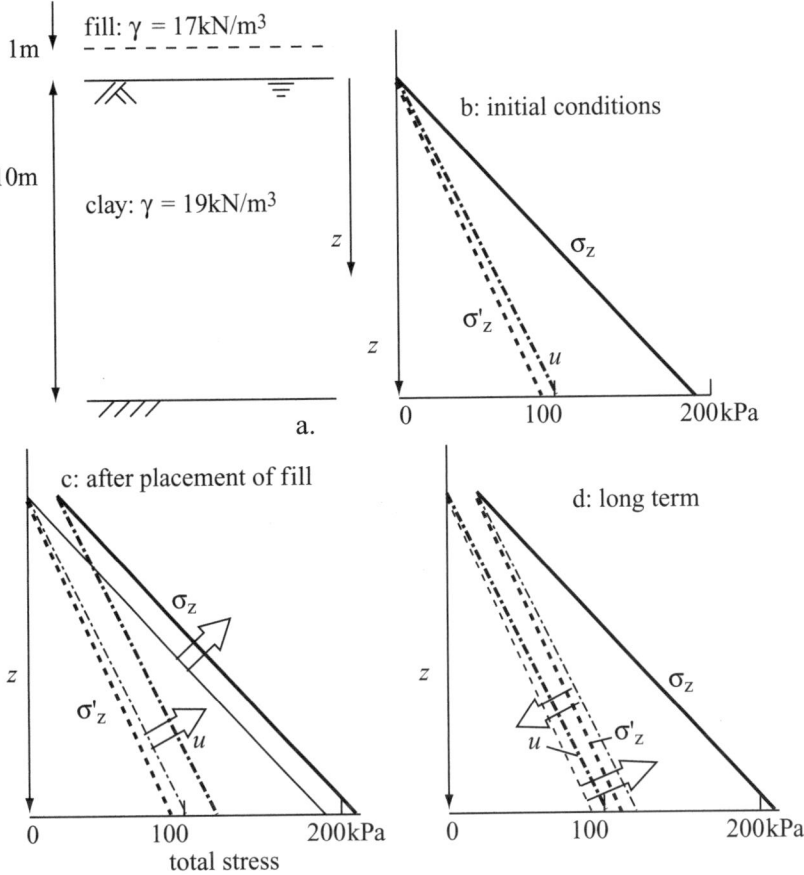

Figure 6.4. Soil conditions for Example 2, Section 6.3.2: distributions of total stress σ_z, effective stress σ'_z and pore pressure u.

6.3.2 Example 2

A bed of clay 10 m thick with the water table at its surface is loaded with a layer of fill 1 m thick over a large site (Fig 6.4). Calculate and sketch the profiles of total vertical stress, effective vertical stress and pore pressure (a) just before the fill has been placed; (b) just after the fill has been placed; and (c) in the long term when pore pressure equilibrium has been re-established with the water table still at the surface of the clay. Calculate the immediate and the long-term settlements of the clay surface. The clay has bulk unit weight $\gamma = 19$ kN/m³ and average one-dimensional stiffness $E_o = 1.5$ MPa. The fill has unit weight $\gamma = 17$ kN/m³.

Before the fill is placed, the vertical total stress is calculated from the bulk unit weight and varies with depth according to the relation:

$$\sigma_z = \gamma z = 19z \text{ kPa}$$

The pore pressure is similarly calculated from the depth below the water table, which is conveniently located at the ground surface:

$$u = \gamma_w z = 9.81z \text{ kPa}$$

The vertical effective stress is just the difference between these two profiles:

$$\sigma'_z = (\gamma - \gamma_w)z = (19 - 9.81)z = 9.19z \text{ kPa}$$

These profiles of stress and pore pressure are shown in Fig. 6.4b.

When the fill is placed, there is an increase of total stress at all depths equal to 17 kPa (1 m of fill of unit weight 17 kN/m³). So, if we continue to measure depths from the surface of the clay (and not from the new ground surface, which would confuse), we can describe the total vertical stress:

$$\sigma_z = 19z + 17 \text{ kPa}$$

The vertical effective stress in the clay is not able to change rapidly so, immediately after the fill has been placed it has the same variation with depth as before:

$$\sigma'_z = 9.19z \text{ kPa}$$

and, as usual, the pore pressure has to take up the difference:

$$u = \sigma_z - \sigma'_z = 9.81z + 17 \text{ kPa}$$

and, of course, the change in pore pressure is 17 kPa at all depths in the clay. These profiles of stress and pore pressure are shown in Fig. 6.4c with the heavy lines; the lighter lines show the distributions of σ_z and u in the original state before the placement of the fill.

In the long term, the pore pressure will be able to re-establish an equilibrium with the unchanged water table:

$$u = 9.81z \text{ kPa}$$

The total stress has not changed:

$$\sigma_z = 19z + 17 \text{ kPa}$$

and the effective stress is once again the difference between total stress and pore pressure:

$$\sigma'_z = \sigma_z - u = 9.19z + 17 \text{ kPa}$$

and now the change in effective stress is 17 kPa at all depths in the clay. These profiles of stress and pore pressure are shown in Fig. 6.4d with the heavy lines; the lighter lines show the distributions of σ'_z and u at the time immediately after the placement of the fill.

We can calculate the settlements from knowledge of the changes in effective stress and from the given one-dimensional stiffness of the clay.[2] Immediately after the fill is placed there is, for this one-dimensional situation, no possibility of change of effective stress and hence no settlement. In the long term, the effective stress

---

[2] A single value of E_o has been quoted for simplicity as the average stiffness relevant to the current effective stress and change in effective stress. A more elaborate calculation of settlement could be made using a stiffness which varies with stress level, as described in Section 4.6.1.

6.3 Worked examples

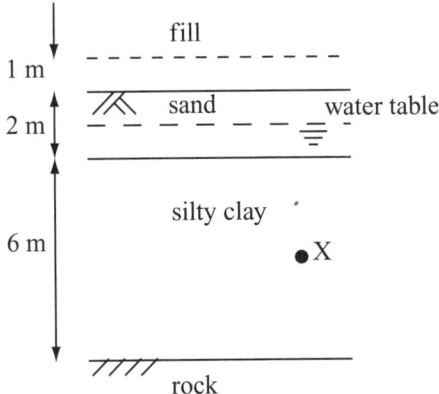

Figure 6.5. Soil conditions for Example 3, Section 6.3.3.

changes by 17 kPa throughout the clay and the vertical strain is simply the ratio of change in effective stress and stiffness:

$$\varepsilon_z = \frac{\Delta\sigma'_z}{E_o} = \frac{17}{1.5 \times 1000} = 1.1\%$$

so that the long-term settlement occurring over the 10 m thick clay layer is $0.011 \times 10 = 0.11$ m.

6.3.3 Example 3

The soils at a certain site consist of 2 m of sand with void ratio $e = 0.6$ overlying 6 m of silty clay with void ratio $e = 1.2$ (Fig 6.5). The silty clay is underlain by rock. The specific gravity of the soil particles of both soils is $G_s = 2.7$. The water table is at a depth of 1 m. Above the water table the sand is dry ($S_r = 0$), and below the water table the sand and silty clay are fully saturated ($S_r = 1$). The site is prepared for construction by the addition of a layer of sandy gravel of thickness 1 m. This material has a void ratio $e = 0.5$ and $G_s = 2.7$. The silty clay has a one-dimensional stiffness $E_o = 1500$ kPa. Calculate the history of stress changes at a point X at the mid-height of the silty clay layer and calculate the long-term settlement of the top of the silty clay, at the interface with the sand.

This example allows us to pull together concepts that we have introduced in several of the preceding chapters. We have to start by converting the information of void ratios and degrees of saturation into densities or unit weights. Then we can analyse the *in-situ* stress and water pressure conditions. Next, we can explore what happens to the stresses and water pressures in the short term and in the long term and, using information about the stiffness of the soils, discover what the eventual settlement will be across the site.

For the sand above the water table, $\rho = G_s \rho_w/(1+e) = 1.69$ Mg/m³; for the sand below the water table, $\rho = (G_s + e)\rho_w/(1+e) = 2.06$ Mg/m³. For the silty clay, $\rho = (G_s + e)\rho_w/(1+e) = 1.77$ Mg/m³.

Point X is at the centre of the silty clay (at a depth below the ground surface of 5 m). Before the site preparation works begin, the total vertical stress is $\sigma_z = (1 \times 1.69 + 1 \times 2.06 + 3 \times 1.77) \times 9.81 = 88.96$ kPa. The pore pressure is calculated from the known level of the water table: $u = 4 \times 1 \times 9.81 = 39.24$ kPa. The effective vertical stress is $\sigma'_z = \sigma_z - u = 49.72$ kPa.

The density of the sandy gravel that is used to provide the fill is $\rho = G_s \rho_w/(1 + e) = 1.8$ Mg/m^3, so that the change in total vertical stress everywhere from the placement of 1 m of this material must be $\Delta \sigma_z = 1 \times 1.8 \times 9.81 = 17.66$ kPa. At all times after the placement of this fill, the total vertical stress at the centre of the silty clay at point X is thus $\sigma_z = 88.96 + 17.66 = 106.62$ kPa. The effective stresses in the silty clay are not able to change until the pore water has been able to flow out of the voids, so the effective stress at point X remains $\sigma'_z = 49.72$ kPa immediately after the placement of the fill. The pore pressure has to take up the difference to satisfy vertical equilibrium: $u = 106.62 - 49.72 = 56.90$ kPa. The deductive sequence of this calculation for time $t = 0$ is thus: total stress → effective stress → pore pressure.

In the long term, pore pressure equilibrium is re-established, with the water table at its original level. The total stresses have not changed but the effective stress has taken up that part of the total stress that was for the time being carried by the pore pressure. Therefore, at point X the total vertical stress after a long time will still be $\sigma_z = 106.62$ kPa. The pore pressure will have reverted to its original value, $u = 39.24$ kPa, and the effective stress is the difference between total stress and pore pressure, $\sigma'_z = \sigma_z - u = 67.38$ kPa. We can check that the increase in effective stress compared with the original value before filling began is equal to the increase in total stress caused by the addition of the layer of fill: $\Delta \sigma'_z = 67.38 - 49.72 = 17.66$ kPa. The deductive sequence of this calculation for time $t \to \infty$ is thus: total stress → pore pressure → effective stress.

This change in vertical effective stress will occur throughout the silty clay layer so, taking an average view of the stress change and of the properties of the clay, the vertical strain in the long term will be: $\varepsilon_z = \Delta \sigma'_z / E_o = 17.66/1500 = 0.012$ or 1.2% throughout the clay layer, which can be converted into a settlement $\varsigma = $ vertical strain × layer thickness $= 0.012 \times 6 = 0.071$ m.

6.4 Summary

Here is a concise list of the key messages from this chapter, which are also encapsulated in the mind map (Fig. 6.6).

1. The Principle of Effective stress tells us that total stress can be shared between the pore pressure in the water in the voids of the soil and the effective stress carried by the soil particles: $\sigma_z = u + \sigma'_z$.
2. Change in effective stress $\Delta \sigma'_z$ implies deformation of the "soil spring". Deformation of the "soil spring" requires movement of the pore fluid from or to the voids around the soil particles. For low-permeability soils saturated with water

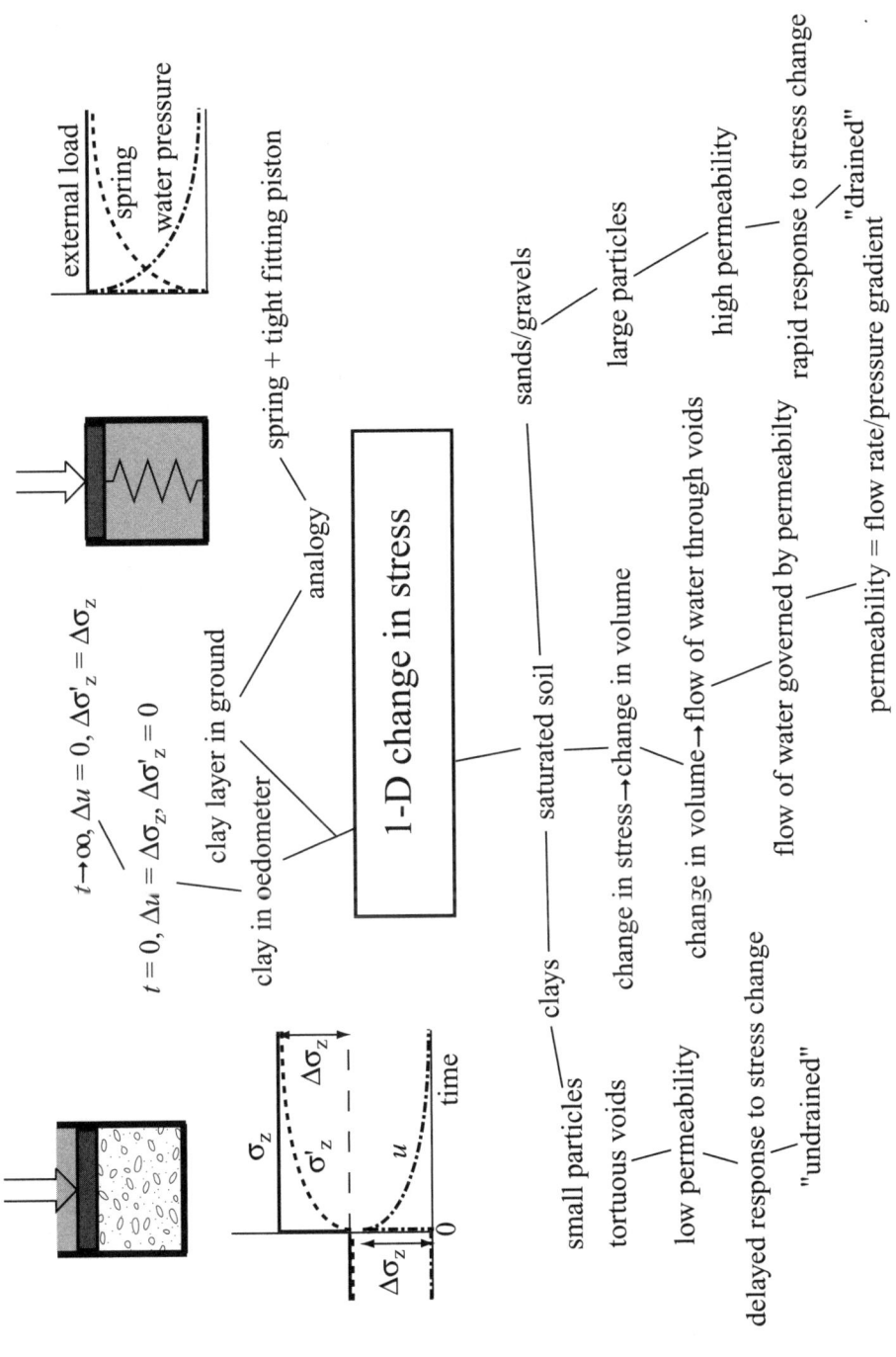

Figure 6.6. Mind map: one-dimensional change in stress.

(which is essentially incompressible), this fluid flow cannot occur rapidly and immediate changes in total stress $\Delta\sigma_z$ become corresponding changes in pore water pressure Δu.

3. At time $t = 0$, $\Delta\sigma'_z = 0$, $\Delta u = \Delta\sigma_z$; as $t \to \infty$, $\Delta\sigma'_z \to \Delta\sigma_z$ and $\Delta u \to 0$.
4. The immediate response is described as undrained and the long-term response as drained.
5. The distinction between undrained and drained response depends on the permeability of the soil, the nature of the loading, and the thickness of the soil layer which needs to drain.

6.5 Exercises: Change in stress

1. A sample of clay is subjected to a total vertical stress $\sigma_z = 150$ kPa. The pore pressure is $u = 10$ kPa. What is the vertical effective stress σ'_z?
2. The total stress on the sample in Question 1 is increased to 200 kPa. What are the new values of u and σ'_z? The one-dimensional stiffness of the clay is $E_o = 1500$ kPa. What is the vertical strain immediately after the vertical stress is increased?
3. The total stress on the sample of clay in Question 2 remains unchanged for a very long time, and the pore pressure eventually returns to its initial value of 10 kPa. What are the long-term values of σ_z, u and σ'_z? What is the eventual vertical strain in the sample?
4. An oedometer sample of clay of diameter 100 mm and thickness 20 mm is in equilibrium under an external load of 1.2 kN with zero pore pressure. Calculate the vertical total stress and vertical effective stress.
5. The load on the sample in Question 4 is increased to 1.8 kN. What are the values of total stress, effective stress and pore pressure immediately after the increase in load? The clay has a stiffness $E_o = 1750$ kPa. What is the immediate change in height of the sample?
6. The load on the sample in Question 5 is maintained at 1.8 kN for a long time. What are the eventual total stress, effective stress and pore pressure, and what are the final vertical strain and the final height of the sample?
7. A bed of clay 10 m thick with the water table always at its surface (saturated unit weight 18 kN/m³) is overlain by a layer of sand 2 m thick (dry unit weight 16.5 kN/m³). Calculate and sketch the variation with depth of total vertical stress, pore pressure and effective vertical stress through these soils.
   The top 1 m of sand is removed. Calculate and sketch the total and effective stress and pore pressure profiles (a) immediately after the sand is removed and (b) in the long term.
   Calculate the immediate and long-term change in thickness of the clay. The one-dimensional confined stiffness is $E_o = 1.7$ MPa.
8. A bed of clay 10 m thick with the water table at its surface is overlain by a layer of sand 2 m thick. The bed of clay is underlain by a second layer of sand. The properties of the soils are the same as in Question 7.

6.5 Exercises: Change in stress

Water is now pumped from the lower layer of sand while the free water table remains at the surface of the clay. The long-term effect is to produce a linear variation of pore pressure from zero at the surface of the clay to only 65 kPa at the base of the clay.

Calculate and sketch the variation of total vertical stress, pore pressure and effective vertical stress in the long term. Calculate the long-term movement of the surface of the clay.

9. At a site where a large factory is to be built, exploratory borings are made, and the information obtained from samples taken from one borehole near the centre of the site reveals 3 m of sand over 4 m of clay underlain by bedrock. The water table is at a depth of 1 m. The sand has a void ratio 0.6 and specific gravity $G_s = 2.7$. It is dry above the water table ($S_r = 0$) and saturated below the water table ($S_r = 1$). It has a high permeability, and offers no resistance to the flow of water.

The clay is saturated with unit weight $\gamma = 20$ kN/m³. It has a low permeability and has a one-dimensional confined stiffness of $E_o = 1.6$ MPa. The rock may be regarded as impermeable.

Calculate and sketch the distribution with depth of total vertical stress σ_z, pore pressure u, and effective vertical stress σ'_z.

If the top 0.5 m of soil were excavated over the entire site, what immediate changes would you expect in any of these pressures or stresses? Assuming that the water table remains unchanged, plot the distributions of total and effective stresses and pore pressure (a) immediately after excavation and (b) in the long term.

What will be the long-term change in thickness of the clay?

10. The soil underlying the city of Venice is a clay layer of thickness about 300 m with an average saturated unit weight $\gamma = 19$ kN/m³. Water has been pumped out of an underlying aquifer for many years so that a stable linear profile of pore pressure has developed from 0 kPa at the ground surface to only 2700 kPa at the base of the clay. Calculate profiles of total and effective stress and pore pressure.

International pressure forces the city to stop pumping from the aquifer and the pore pressure eventually returns to the equilibrium hydrostatic value throughout the clay. Estimate the resulting movement of the city of Venice. The average stiffness of the clay is $E_o = 108$ MPa.

7 Consolidation

7.1 Introduction

In previous chapters, we have noted that soils of low permeability – typically, clayey soils – will not be able to respond rapidly to changes in stresses which require deformation, which in turn implies, for our one-dimensional systems, change in volume. We suggested in Section 6.2 that there would be a transient process between time $t = 0$ and time $t \to \infty$ during which the temporary disequilibrium of pore pressures would disappear as deformation of the soil spring permitted stresses to be transferred from pore pressure to effective stresses supported by the soil itself. The detail of this process was not considered: we merely looked at the two extremes of the immediate aftermath of the change in external stress (at $t = 0$) and the eventual equilibrium at infinite time ($t \to \infty$). In this chapter, we will explore various aspects of the analysis of the transient process between these two extremes. The transient process is known as *consolidation*.

We will first produce an approximate solution which forces us to concentrate on the important physical aspects of the problem. However, the governing differential equation – the *diffusion equation* – is one that is common to many problems which involve gradient driven flow: heat flows down a temperature gradient, pollution flows down a concentration gradient, pore water flows down a gradient of total head (Sections 5.2, 5.4). The diffusion equation is common to a range of problems, and the development and solution of the governing equation provides a fairly straightforward example of the use of mathematics in the service of the mechanics of soils. It is instructive to see how the exact solutions and the approximate solutions compare for different regimes of consolidation.

It is helpful to separate that part of the pore pressure which is associated with a stable flow regime (which might be a regime of no flow or a regime of steady flow) from that part of the pore pressure which forms part of the transient response to some perturbation. We will call this second part the *excess* pore pressure, which will become zero in the steady long-term state. We will use the symbol u' to denote this excess pore pressure. In the language of Chapter 5 and the analysis of one-dimensional flow, the first part of the pore pressure is associated with either

7.1 Introduction

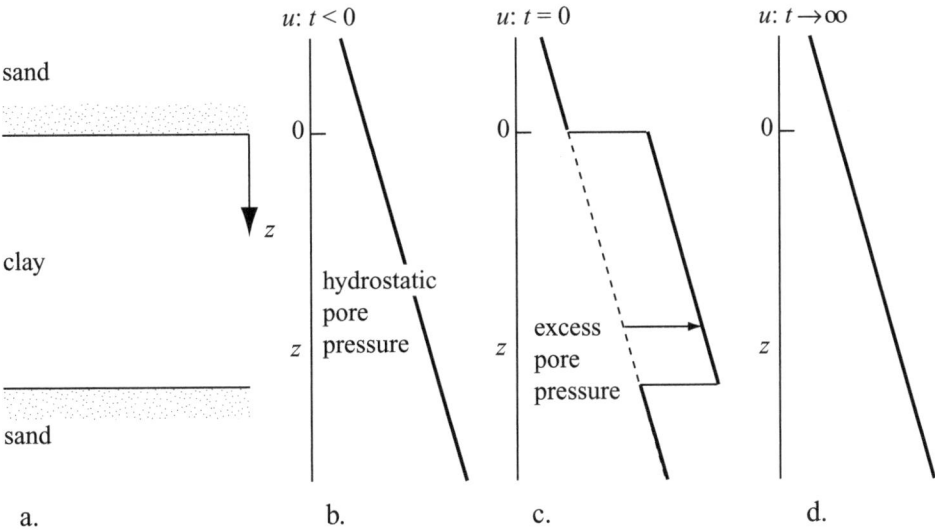

Figure 7.1. (a) Layer of clay between layers of sand: problem definition; (b) pore pressure at time $t < 0$; (c) pore pressure at time $t = 0$; (d) pore pressure at time $t \to \infty$.

constant total head H or linearly varying total head throughout the soil layer of interest. Thus, Figs 7.1 and 7.2 are equivalent: the first shows the distributions of pore pressure; the second shows the distributions of excess pore pressure. Figures 7.1b, d and 7.2b, d show the steady state distributions, which are here assumed hydrostatic through all soil layers, at times $t < 0$ and $t \to \infty$; Figs 7.1c and 7.2c show the distributions at time $t = 0$ just after some surcharge has been placed which can only be supported in the impermeable clay layer by an increase in pore pressure. This

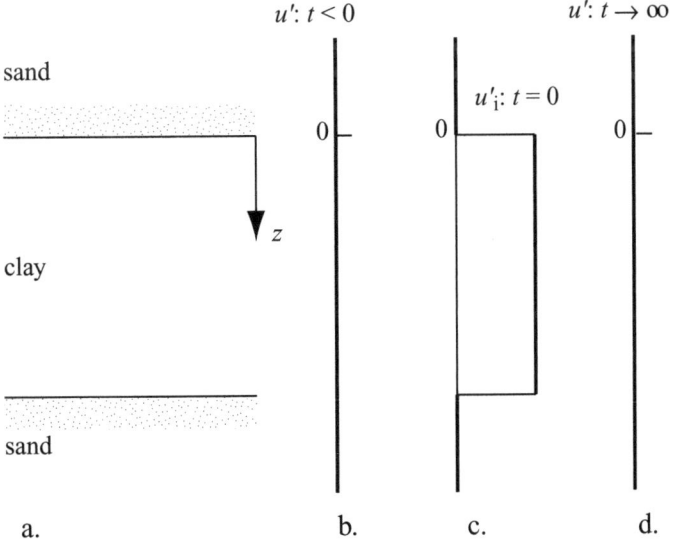

Figure 7.2. (a) Layer of clay between layers of sand: problem definition; (b) excess pore pressure at time $t < 0$; (c) excess pore pressure at time $t = 0$; (d) excess pore pressure at time $t \to \infty$ (compare Fig. 7.1).

increase is the excess pore pressure, the process of whose transient dissipation we will analyse in this chapter.

7.2 Describing the problem

The ground is sitting quite happily, with the pore water quietly occupying the spaces between the soil particles and with the pressure in the water corresponding to some steady groundwater regime (no flow or steady flow), and with the soil particles themselves supporting an appropriate part of the external loads to maintain equilibrium in accordance with the Principal of Effective stress. This is the description that applies for all times $t < 0$.

But suddenly, at time $t = 0$, there is a change and this idyll is destroyed. Some external loading is increased – for example, by the placing of a layer of fill as in Section 6.3.2, which we assume can happen in the relative twinkling of an eye. We suppose that we are concerned with a somewhat impermeable layer of clay, as in that example. We know what happens: the clay is incapable of rapid deformation because of its low permeability; therefore there is no possibility of any change in the effective stress that is carried by the soil particles. The soil spring (by analogy with Fig. 6.1) is unable to compress, and therefore the pore pressure has to increase throughout the layer of clay in order to support the changed external loads and still maintain equilibrium. The pore pressure takes up the increase in total stress while the effective stress remains unchanged. That is the conclusion we reached in Chapter 6.

Let us suppose that our layer of clay is sandwiched between more permeable layers of sand (Fig. 7.2). The sand is able to respond immediately to the change in external conditions, the necessary flow of pore water is able to occur allowing the sand to compress (although the compression of the sand will probably be insignificant compared with the eventual compression of the clay) and the effective stress supported by the soil particles is able to increase. The pore pressure in the sand remains unchanged and hence the excess pore pressure in the sand is zero (Figs 7.1c, 7.2c). At the boundary between the two soils – sand and clay – at $z = 0$ (Fig. 7.2), we have two values of excess pore pressure. Just above the boundary, in the sand, $z < 0$, the excess pore pressure is zero (the pore water is in hydraulic equilibrium with its surroundings) but just below the boundary, in the clay, $z > 0$, the excess pore pressure is equal to the change in external loading. At this boundary, therefore, the gradient of the excess pore pressure is infinite, $du'/dz = \infty$, but on either side of the boundary the gradient of excess pore pressure is zero. We will focus initially only on the upper boundary of the clay, at $z = 0$; the route to the analysis of the lower boundary will become evident later.

We know, from Chapter 5, that flow of pore water will occur whenever there is a gradient of total head or excess pore water pressure. Our simple model of seepage is enshrined in Darcy's Law (5.23):

$$\bar{v} = -ki = -\frac{k}{\gamma_w}\frac{du'}{dz} \tag{7.1}$$

7.2 Describing the problem

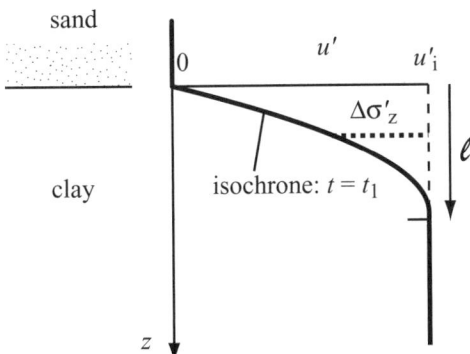

Figure 7.3. Consolidation front penetrating distance ℓ into layer of clay at time $t = t_1$, defining isochrone of excess pore pressure for $t = t_1$.

where $\bar{v}$ is the flow velocity averaged over the cross-section of the soil, k is the coefficient of permeability, and i is the hydraulic gradient which describes the gradient of total head within the soil. In analysing transient consolidation, it is helpful to work with excess pore pressure u' rather than head: pressure and head are linked through the unit weight $\gamma_w = \rho_w g$ of the water. The negative sign in (7.1) is included as a formal indication that the water flows down the gradient of excess pore pressure or total head. It is convenient to suppose for simplicity here that there is no flow except that resulting from the change in external stress, so that in the long term, once hydraulic equilibrium has been re-established at $t \to \infty$, $i = (1/\gamma_w) du'/dz = dH/dz = 0$. Everything that we do is concerned with the departure from this eventual steady state.

Darcy's Law tells us that, for an instant, there will be an infinite flow rate at the boundary of the clay; but it also tells us that within the clay layer there will be no flow because there is no gradient of excess pore pressure: at $z = 0$, $i = \infty$; for $z > 0$, $i = 0$ (Fig. 7.2b). The clay within the layer is quite oblivious to the fact that there is a boundary at $z = 0$, from which drainage has started to occur. It will take time for this knowledge to penetrate.

Flow implies movement of pore water and basic ideas of conservation of mass must apply. In fact, since we are assuming that both the water and the soil particles are incompressible, conservation of mass implies conservation of volume. The soil skeleton which describes the locations of the soil particles is not incompressible: as water flows out of or into the voids, the volume defined by the soil skeleton and occupied by the soil particles will change. For a saturated soil, this is the only way in which volume change can occur. Conservation of volume then tells us that, if water is flowing out of the clay and into the sand, then the volume of the clay (or, better, the volume occupied by the clay) must be decreasing. The volume is decreasing because of the transfer of total stress from pore water pressure to effective stress. The rate at which the clay is compressing must exactly balance the rate at which water is flowing out of the clay.

The concept of infinity is difficult to handle, so let us consider a time $t_1 > 0$ which is typical of the process of gradual penetration of the *consolidation front* into the clay (Fig. 7.3). There will be some variation of excess pore pressure through the clay, which we can describe using an *isochrone*.[1] At the boundary of the clay, $z = 0$, the slope of the pore pressure isochrone defines the rate of flow of water out of the clay. For cross-sectional area A of the clay, the volumetric rate of flow is q:

$$q = A \frac{k}{\gamma_w} \left[\frac{du'}{dz}\right]_{z=0} \tag{7.2}$$

This flow rate must balance the rate of change in volume of the clay resulting from the change in effective stress. If the consolidation front has penetrated a distance ℓ (Fig. 7.3), then at each depth $0 < z < \ell$ the strain in the clay is related to the change in effective stress $\Delta\sigma'_z$ scaled by the one-dimensional stiffness of the clay, E_o. The change in effective stress is the separation of the initial pore pressure isochrone for $t = 0$ (for which the excess pore pressure is u'_i throughout the layer) and the current pore pressure isochrone: $\Delta\sigma'_z = u'_i - u'$ (Fig. 7.3). The vertical strain at depth z is $\Delta\varepsilon_z = \Delta\sigma'_z/E_o$ and the accumulated settlement of the layer is:

$$\varsigma = \int_0^\ell \Delta\varepsilon_z dz = \frac{1}{E_o}\int_0^\ell (u'_i - u')dz \tag{7.3}$$

Conservation of volume requires that the rate of flow of water out of the clay must match the rate of settlement of the clay layer:

$$q = A\frac{d\varsigma}{dt} \tag{7.4}$$

or, with (7.2):

$$\frac{k}{\gamma_w}\left[\frac{du'}{dz}\right]_{z=0} = \frac{d}{dt}\left[\frac{1}{E_o}\int_0^\ell (u'_i - u')dz\right] \tag{7.5}$$

This looks rather a tricky equation to solve but it becomes tractable if we make some reasonable assumption about the shape of the pore pressure isochrone at time $t = t_1$. We then have immediate access to both the exit gradient controlling the flow rate at $z = 0$ and the average effective stress change from the integral of the area between the isochrones for $t = 0$ and $t = t_1$.

7.3 Parabolic isochrones

Let us assume that at all times the pore pressure isochrones have a common geometric shape. The assumption of an appropriate plausible *shape function* for the variation of some quantity within a defined region is a standard approximation technique which will be encountered in other engineering applications. In principle, we could assume any shape. It turns out that the mathematics becomes particularly simple if we assume that the isochrones are parabolic.[2] The excess pore pressure is assumed to vary parabolically with distance from the drainage boundary (Fig. 7.3). The effective stresses are given by the difference between the applied total stress and this

---

[1] *isochrone*: from Greek "equal time".
[2] Schofield, A.N. and Wroth, C.P. (1968) *Critical state soil mechanics*, McGraw-Hill, London.

7.3 Parabolic isochrones

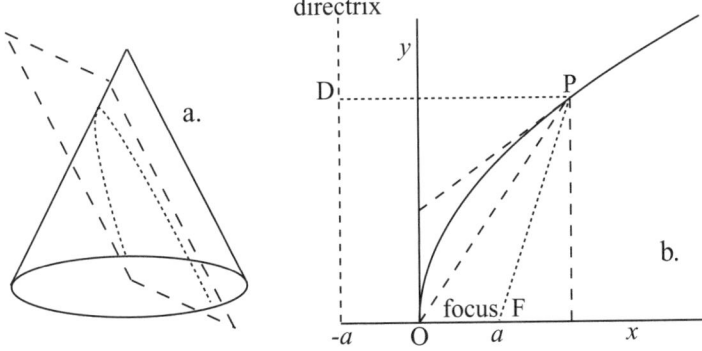

Figure 7.4. Geometry of the parabola: (a) conic section; (b) focus and directrix.

pore pressure so that the volume change of the clay is linked with the area above the parabolic isochrone scaled with the one-dimensional stiffness, E_o. The rate at which water flows out of the clay is controlled by the slope of the parabolic isochrone at the drainage boundary. The problem enshrined in (7.5) then reduces to a simple differential equation, deduced from the geometry of the parabola, linking the rate of change of volume of the soil with the rate at which water flows out of the soil.

The parabola is a curve obtained by cutting a cone parallel to its sloping side (Fig. 7.4a). Its equation can be derived by noting that it is the locus of points which are equidistant from a point – the focus, F: $(x, y) = (a, 0)$ – and a straight line – the directrix: $x = -a$ (Fig. 7.4b). For any point P, PF = PD, or:

$$\sqrt{(x-a)^2 + y^2} = a + x \tag{7.6}$$

or, squaring both sides and rearranging:

$$y^2 = 4ax \quad \text{or} \quad y = 2\sqrt{ax} \tag{7.7}$$

Differentiating this equation, we can show that the slope of the parabola at P is:

$$\frac{dy}{dx} = \frac{y}{2x} \tag{7.8}$$

which is half the slope of the line from the origin O to P, the diagonal of the enclosing rectangle y/x.

Integrating (7.7), we can find the area under the parabola from O to P:

$$\text{area} = \int_0^x 2\sqrt{ax}\,dx = \frac{4}{3}x\sqrt{ax} = \frac{2}{3}xy \tag{7.9}$$

and the area is two thirds of the enclosing rectangle. The elements of the geometry of the parabola that we need to use in our interpretation of the process of consolidation are thus rather simple (Fig. 7.5).

We can now tackle our governing equation (7.5). From (7.8), the left hand side can be written:

$$\frac{k}{\gamma_w}\left[\frac{du'}{dz}\right]_{z=0} = \frac{2ku'_i}{\gamma_w \ell} \tag{7.10}$$

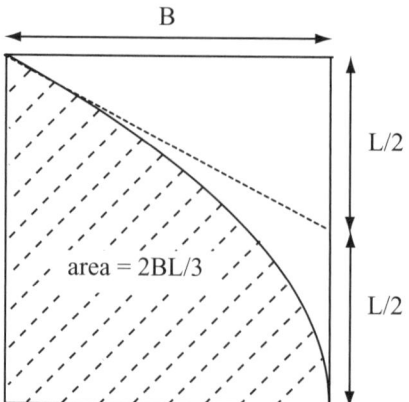

Figure 7.5. Geometry of parabolic isochrone.

and from (7.9) the right hand side can be written:

$$\frac{d}{dt}\left[\frac{1}{E_o}\int_0^\ell (u'_i - u')dz\right] = \frac{d}{dt}\left[\frac{u'_i \ell}{3E_o}\right] = \frac{u'_i}{3E_o}\frac{d\ell}{dt} \quad (7.11)$$

Equating these, we find the first order ordinary differential equation:

$$\frac{u'_i}{3E_o}\frac{d\ell}{dt} = \frac{2ku'_i}{\gamma_w \ell} \quad (7.12)$$

or:

$$2\ell d\ell = 12\frac{kE_o}{\gamma_w}dt \quad (7.13)$$

which we can integrate to give:

$$\ell = \sqrt{12\frac{kE_o}{\gamma_w}t} \quad (7.14)$$

This is an expression for the distance ℓ that the boundary disturbance (the sensation of the pore pressure discontinuity at $z = 0$ and $t = 0$) is able to penetrate in time t. We note, firstly, that this expression is independent of u'_i – the magnitude of the pore pressure disturbance is not important. We note, secondly, that the location of the consolidation front, the value of ℓ, is proportional to the square root of time. And we note, thirdly, that the soil property controlling the process is a composite quantity which is called the *coefficient of consolidation* c_v:

$$c_v = \frac{kE_o}{\gamma_w} \quad (7.15)$$

The coefficient of consolidation has dimensions of (length^2/time) and should be quoted in standard SI units of metres^2/second. The range of values of c_v is high and the values that are likely to be encountered for geotechnical problem where effects of consolidation are a crucial design factor tend to be quite low, so that non-standard units are often used. This merely calls for care in ensuring consistency

7.3 Parabolic isochrones

Table 7.1. *Typical values of coefficient of consolidation, c_v.*

	Quoted units	m^2/s
8×10^{-2}	m^2/day	0.93×10^{-6}
60×10^{-6}	m^2/min	1.00×10^{-6}
0.01	ha/year	3.17×10^{-6}
0.4	m^2/year	12.7×10^{-9}
1.3	mm^2/min	21.7×10^{-9}
0.1	m^2/month	38.6×10^{-9}
8×10^{-4}	cm^2/s	80.0×10^{-9}

in any calculation. A quick trawl of a random selection of literature produces the values and units shown in Table 7.1 for the coefficient of consolidation of clays. They have been converted to standard units in the final column.

Knowing how far the sensation of consolidation has penetrated is part of the story; it is useful also to know how much settlement has occurred and how rapidly it is occurring. Settlement was given by (7.3) and this becomes:

$$\varsigma = \frac{1}{E_o}\int_0^\ell (u_i' - u')dz = \frac{u_i'\ell}{3E_o} = \frac{u_i'}{3E_o}\sqrt{12\frac{kE_o}{\gamma_w}t} = u_i'\sqrt{\frac{4kt}{3E_o\gamma_w}} \qquad (7.16)$$

The settlement also varies with the square root of time but is, not surprisingly, proportional to the pore pressure discontinuity.

There is no restriction on the thickness of the clay layer to which this analysis can be applied, but it will cease to be valid once the entire clay layer is aware of the occurrence of drainage. We can imagine two simple examples of layers of finite thickness: a layer of thickness L sandwiched between a layer of permeable sand and a layer of impermeable rock (Fig. 7.6a); and a layer of thickness $2L$ sandwiched between two layers of permeable sand (Fig. 7.6b). In fact, from symmetry, the two configurations are equivalent. The boundary drainage in Fig. 7.6b creeps in simultaneously from top and bottom and the whole layer is consolidating once $\ell = L$. From symmetry, there can be no flow across the mid-height of the layer at distance L from either boundary, and hence $du'/dz = 0$ at $z = L$ for $0 < t < \infty$. This plane of symmetry might just as well be an impermeable plane. The boundary drainage in Fig. 7.6a creeps in from the top boundary only and the whole layer is consolidating once $\ell = L$. The real impermeability of the lower boundary means that the isochrones must always meet it orthogonally and $du'/dz = 0$ at $z = L$ for $0 < t < \infty$.

We will consider the layer of thickness L, noting that the rate of consolidation will be the same for both layers but that the settlements will be doubled for the layer of thickness $2L$. The first stage of the consolidation ends when $\ell = L$ at $t = t_A = L^2/12c_v$. At this time, the excess pore pressure at depth $z = L$ is still just equal to the original value u_i'. In the second stage of the consolidation, the entire layer is participating, and the pore pressure at the base of the layer falls progressively to zero. The typical isochrone for some intermediate time $t_A \le t < \infty$ is shown in Fig. 7.7 with pore pressure u_L' at the lower boundary. Exactly the same physical

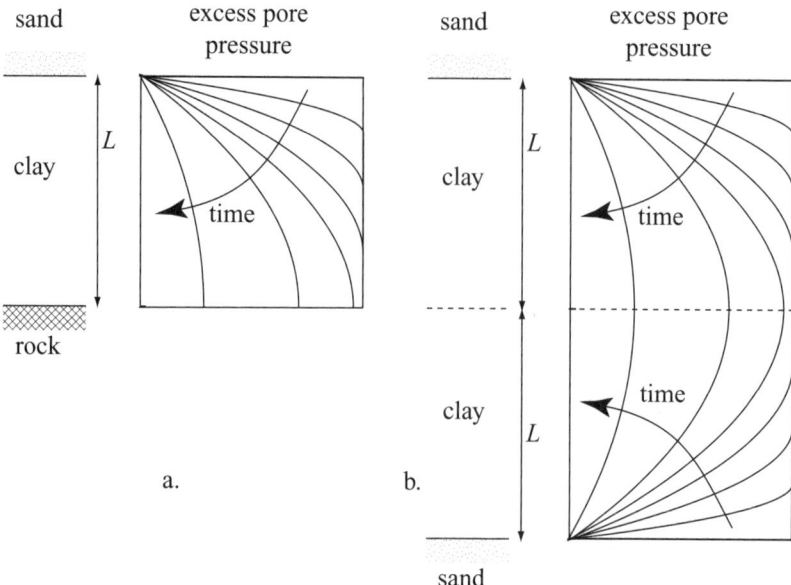

Figure 7.6. (a) Clay layer of thickness L between layer of permeable sand and impermeable rock; and (b) clay layer of thickness $2L$ between two layers of permeable sand.

reasoning applies, equating the flow out of the layer (from the hydraulic gradient at the top boundary) with the rate of change of settlement (from the area above the current isochrone).

We return to our governing equation (7.5). From (7.8) the left hand side can be written:

$$\frac{k}{\gamma_w}\left[\frac{du'}{dz}\right]_{z=0} = \frac{2ku'_L}{\gamma_w L} \qquad (7.17)$$

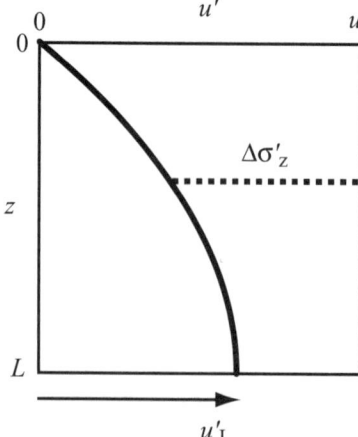

Figure 7.7. One-dimensional consolidation, parabolic isochrones, stage 2: fall in pore pressure at undrained boundary.

7.3 Parabolic isochrones

and from (7.9) the right hand side can be written:

$$\frac{d}{dt}\left[\frac{1}{E_o}\int_0^L (u_i' - u')dz\right] = \frac{d}{dt}\left[\frac{L}{3E_o}u_L' + \frac{L}{E_o}(u_i' - u_L')\right] = -\frac{2L}{3E_o}\frac{du_L'}{dt} \quad (7.18)$$

Equating these, we find the first order ordinary differential equation:

$$-\frac{2L}{3E_o}\frac{du_L'}{dt} = \frac{2ku_L'}{\gamma_w L} \quad (7.19)$$

or:

$$\frac{du_L'}{u_L'} = -\frac{3kE_o}{\gamma_w L^2}dt \quad (7.20)$$

with the solution:

$$u_L' = u_i' \exp\left(-3c_v(t - t_A)/L^2\right) \quad (7.21)$$

since $u_L' = u_i'$ at $t = t_A$. The settlement is:

$$\varsigma = \frac{1}{E_o}\int_0^L (u_i' - u')dz = \frac{L}{3E_o}u_L' + \frac{L}{E_o}(u_i' - u_L')$$
$$= \frac{Lu_i'}{3E_o}\left[3 - 2\exp\left(-3c_v(t - t_A)/L^2\right)\right] \quad (7.22)$$

We have presented all these results in terms of the several controlling dimensions and variables and soil properties, but the applicability of the expressions for penetration time and settlement can be broadened by presenting the same results in dimensionless form. It is logical to normalise lengths by the thickness L of the clay layer, so we write $\tilde{Z} = z/L$ and $\tilde{L} = \ell/L$. The coefficient of consolidation c_v has dimensions of length^2/time, so we can define a dimensionless time $\tilde{T} = c_v t/L^2$. The excess pore pressure can be normalised by its initial value to give a dimensionless pressure, $\tilde{U} = u'/u_i'$. The eventual, long-term settlement ς_∞ of the layer results from the complete transfer of the initial excess pore pressure to effective stress so that the strain is $\varepsilon_z = u_i'/E_o$ throughout. The final settlement is $\varsigma_\infty = u_i' L/E_o$. We can define a degree of consolidation $\tilde{S} = \varsigma/\varsigma_\infty$ as an indication of the progress of the settlement of the layer towards this long-term goal. A summary of the two stages of the process of consolidation, described in terms of these dimensionless variables, is shown in Fig. 7.8.

During the first stage of consolidation (Fig. 7.8a), the dimensionless depth of penetration of the consolidation front is $\tilde{L}$. Rewriting (7.14):

$$\tilde{L} = \frac{\ell}{L} = \sqrt{12\tilde{T}} \quad (7.23)$$

and this stage ends when $\tilde{L} = 1$ and $\tilde{T} = 1/12$. During this stage the normalised settlement, the degree of consolidation, is $\tilde{S}$. Rewriting (7.16):

$$\tilde{S} = \varsigma/\varsigma_\infty = \frac{u_i'\ell}{3E_o}\frac{E_o}{u_i' L} = \sqrt{\frac{4\tilde{T}}{3}} \quad (7.24)$$

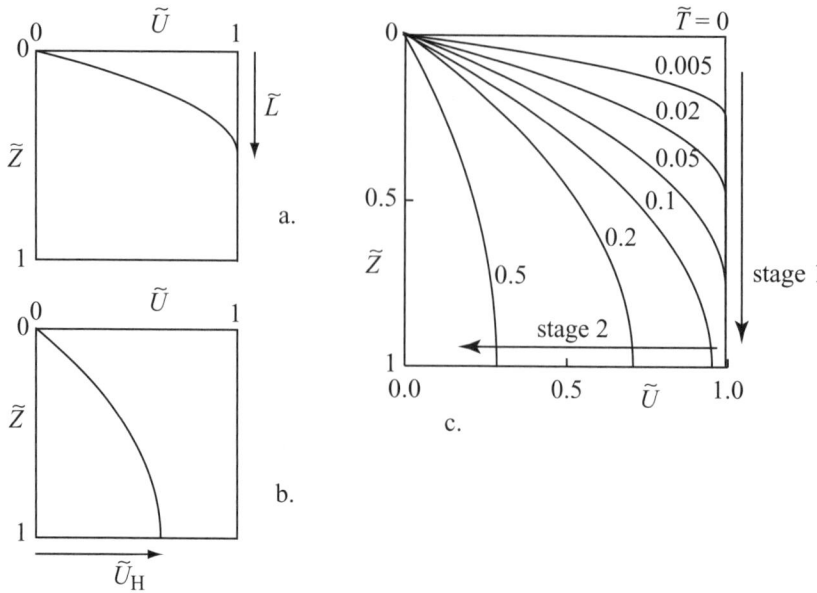

Figure 7.8. One-dimensional consolidation, parabolic isochrones, (a) stage 1: propagation of boundary drainage into layer; (b) stage 2: fall in pore pressure at undrained boundary; (c) isochrones for combined solution.

At the end of the first stage of consolidation, $\tilde{T} = \tilde{T}_A = 1/12$ and $\tilde{S} = 1/3$, which corresponds with our expectation that the area above the isochrone which just reaches the depth L is one third of the full area beneath the initial isochrone, and hence one third of the eventual effective stress change has occurred at this time.

During the second stage of consolidation (Fig. 7.8b), the dimensionless pore pressure at the base of the layer, $\tilde{U}_L = u'_L/u'_i$, at depth $\tilde{Z} = z/L = 1$, varies with dimensionless time $\tilde{T}$, as deduced from (7.21):

$$\tilde{U}_L = \exp\left[-3(\tilde{T} - 1/12)\right] \quad (7.25)$$

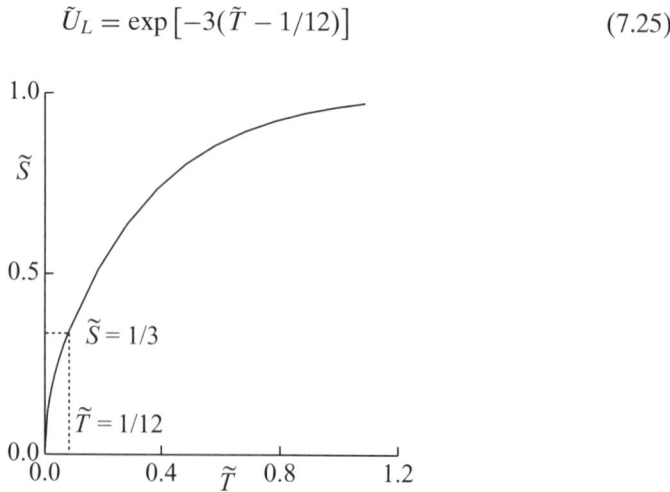

Figure 7.9. Parabolic isochrones, degree of consolidation $\tilde{S}$ and time $\tilde{T}$.

7.4 Worked examples

Table 7.2. *Parabolic isochrones, dimensionless progress of consolidation.*

Degree of consolidation $\tilde{S}$	Dimensionless time $\tilde{T}$	Degree of consolidation $\tilde{S}$	Dimensionless time $\tilde{T}$
0	0		
0.1	0.0075	0.6	0.2536
0.2	0.03	0.7	0.3495
0.3	0.0675	0.8	0.4846
0.33	0.0833	0.9	0.7157
0.4	0.1184	0.95	0.9467
0.5	0.1792	0.99	1.4832

and the dimensionless settlement, $\tilde{S} = \varsigma/\varsigma_\infty$, from (7.22), is:

$$\tilde{S} = 1 - \frac{2}{3}\exp\left[-3(\tilde{T} - 1/12)\right] \tag{7.26}$$

The combination of the two stages of consolidation is shown in Fig. 7.9, and a table of values of dimensionless settlement $\tilde{S}$ and dimensionless time $\tilde{T}$ is given in Table 7.2.

We conclude that we can capture the essence of the consolidation problem by standing one step back from the exact equation and adopting a simpler mathematical description – imposing a particular parabolic mode shape. The controlling physical principles are of course retained but they are applied to the complete system – the complete clay layer – rather than to the individual "sub-layers". The overall physical process can then be followed rather clearly.

7.4 Worked examples

7.4.1 Example 1: Determination of coefficient of consolidation

The determination of the coefficient of consolidation, c_v, from the results of a laboratory oedometer test requires the fitting of the theoretical expressions for the development of settlement with time to the experimental data. A typical strategy for the determination of c_v is illustrated in Fig. 7.10.

During the initial stage of consolidation, while the consolidation front is penetrating into the clay, the settlement varies with the square root of time. We can fit a straight line to this initial section. We can then draw another line (or lines) at a different slope to intersect the second stage of the consolidation process, and the intersection point enables us to match the experiment and the analytical model. Figure 7.10 shows lines drawn at 1.05, 1.1 and 1.2 times the initial gradient. The dimensionless coordinates of the intersection points are given in Table 7.3.

Figure 7.11 applies this process to data obtained from an oedometer test on reconstituted Gault clay following a change of vertical stress, σ_z, from 200 to 400 kPa. The oedometer sample had an initial height of 15.56 mm and it was

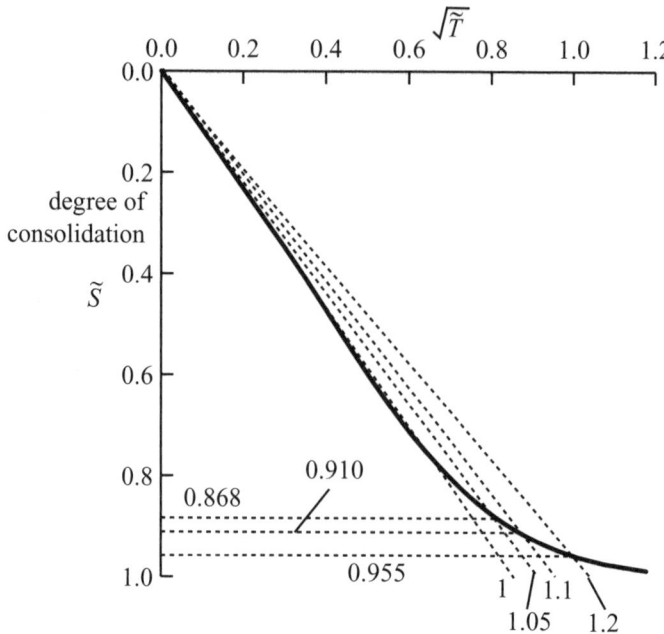

Figure 7.10. Determination of coefficient of consolidation from laboratory oedometer data.

drained from both top and bottom surfaces. The data have been plotted as settlement (in millimetres) against the square root of time (in seconds). The initial stage is fitted with a straight line. A line with a slope 1.1 times this gradient intersects the experimental curve at $\varsigma = 0.815$ mm and $\sqrt{t} = 48.85$ s$^{1/2}$. From Table 7.3, this corresponds to $\tilde{S} = 0.910$, so we can estimate the final settlement, $\varsigma_\infty = 0.815/0.910 = 0.896$ mm, and the average sample height over the increment $2L = 15.11$ mm. The one-dimensional stiffness of the clay would thus be $E_o = \Delta\sigma_z/(\varsigma_\infty/2L) = 200/(0.896/15.11) = 3373$ kPa. Also from Table 7.3, the intersection point corresponds to $\tilde{T} = c_v t/L^2 = 0.752$. We have $t = 48.85^2 = 2386$ s and $L = 7.56$ mm, so that $c_v = 0.752 \times (7.56 \times 10^{-3})^2/2386 = 18 \times 10^{-9}$ m^2/s. Knowing the one-dimensional stiffness E_o and the coefficient of consolidation c_v, we can deduce the permeability of the clay, $k = \gamma_w c_v/E_o = 9.81 \times 18 \times 10^{-9}/3373 = 52.3 \times 10^{-12}$ m/s.

Alternatively, a line with slope 1.2 times the initial gradient intersects the experimental curve at $\varsigma = 0.892$ mm and $\sqrt{t} = 58.30$ s$^{1/2}$ ($t = 3398$ s). From Table 7.3 this corresponds to $\tilde{S} = 0.955$ and $\tilde{T} = 0.986$. We can estimate the final

Table 7.3. Coordinates of intersection points (Fig. 7.10).

Slope	$\tilde{S}$	$\tilde{T}$
1.05	0.868	0.623
1.10	0.910	0.752
1.20	0.955	0.986

7.4 Worked examples

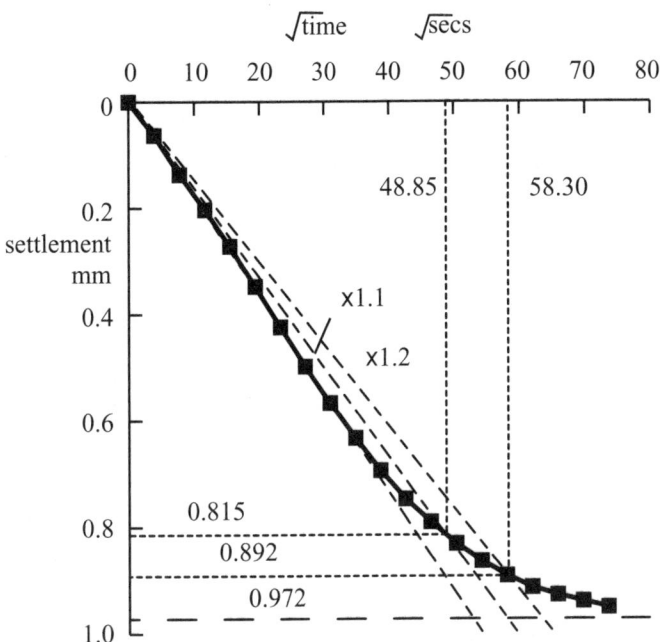

Figure 7.11. Fitting analytical model to laboratory oedometer data (Example Section 7.4.1).

settlement $\varsigma_\infty = 0.892/0.955 = 0.934$ mm, average sample height over the increment $2L = 15.09$ mm, and one-dimensional stiffness $E_o = 200/(0.934/15.09) = 3231$ kPa. Then the coefficient of consolidation $c_v = 0.986 \times (7.55 \times 10^{-3})^2/3398 = 16.5 \times 10^{-9}$ m²/s, and the permeability of the clay $k = \gamma_w c_v / E_o = 9.81 \times 16.5 \times 10^{-9}/3231 = 50.2 \times 10^{-12}$ m/s.

Figure 7.11 shows the actual settlement of $\varsigma = 0.972$ mm reached after the stress of 400 kPa had been left on the sample for about 24 hours (corresponding to an average sample height of 15.08 mm). The actual stiffness over the increment was therefore $E_o = 200/(0.972/15.08) = 3101$ kPa. The clay reveals some small effects of *creep*, continuing compression at constant effective stress, also known as *secondary consolidation*, leading to strains which continue with time at a decreasing rate. It is good practice to use some standard interval for the loading of the oedometer sample (for example, 1 day) so that the effects of creep may be somewhat mitigated. The modelling of time effects such as creep is beyond the scope of this book.

The soil in the oedometer will be aware of the general governing physical rules (conservation of mass, Darcy's Law for gradient driven flow) but may not be so confident of the other assumptions that underpin our analysis (parabolic isochrones, values of soil permeability k and stiffness E_o which remain constant during the process of consolidation). The differences between the estimates of soil properties made using the two fitting points are a reminder that we should not expect that the analytical model and the experimental reality should match each other exactly.

Given oedometer data of the general type shown in Fig. 7.11, we can estimate consolidation settlements and times for real prototype loading events in the field by direct scaling of our laboratory test measurements without having to go through the deduced values of soil properties. Suppose that the clay being tested in the laboratory giving the data in Fig. 7.11 comes from a layer of clay 5 m thick, drained at top and bottom, which is expected to be subjected to a long-term increase in effective stress of 200 kPa. The eventual strain in the prototype will be the same as in the oedometer, $\Delta\varepsilon_z = \varsigma_\infty/2L$. The eventual settlement will be the oedometer settlement scaled by the ratio of thicknesses of the prototype and oedometer clay layers: $\varsigma = 0.972 \times 5/(15.08 \times 10^{-3}) = 322$ mm. From Table 7.2, the dimensionless time for 90% of this settlement to occur is $\tilde{T} = 7157$ which equates to $t = 0.7157 \times (7.55 \times 10^{-3})^2/16.5 \times 10^{-9} = 2467$ s or 41.1 min in the laboratory. The time for consolidation processes scales with the square of the dimension L ($\tilde{T} = c_v t/L^2$). Hence, the time for 90% of the settlement to occur in the prototype will be $t_{90} = 2467 \times (2.5/(7.55 \times 10^{-3}))^2 = 270 \times 10^6$ s $= 8.58$ years. In practice, this might well be seen as rather slow so that some measures would have to be taken to speed the process, which usually means somehow increasing the permeability by installing drains through the clay layer, as described in the next example.

7.4.2 Example 2

We can return to the worked examples from Chapter 6 in order to put some indication of time and rate of settlement into the calculations. The site conditions for the example in Section 6.3.2 are repeated in Fig. 7.12. A bed of clay 10 m thick with the water table at its surface is loaded with a layer of fill 1 m thick over a large site. The clay has bulk unit weight $\gamma = 19$ kN/m^3 and average one dimensional stiffness $E_o = 1.5$ MPa. The fill has unit weight $\gamma = 17$ kN/m^3. We need to add some indication of the permeability of the clay, $k = 10^{-9}$ m/s.

The stress applied by the fill is 17 kPa. We calculated the long term settlement of the clay surface to be 0.11 m. From the given one-dimensional stiffness and permeability, we can calculate the coefficient of consolidation to be $c_v = kE_o/\gamma_w = 10^{-9} \times 1.5 \times 10^3/9.81 = 152.9 \times 10^{-9}$ m^2/s. From Table 7.2, the dimensionless time for 90% consolidation is $\tilde{T} = c_v t/L^2 = 0.7157$. For our site, the time for 90% of the settlement to occur will be $t_{90} = 0.7157 \times 10^2/152.9 \times 10^{-9} = 0.468 \times 10^9$ s $= 14.84$ years, and after this time the settlement will be $\varsigma = 0.9 \times 0.11 = 0.1$ m.

From the definition of our dimensionless time, we can see that we can speed the process of consolidation either by increasing the permeability k or by reducing the thickness of the clay layer L. Neither of these seems obviously straightforward. However, if we think of L not so much as the thickness of the clay layer as the maximum distance in the clay from the drainage boundary, then the challenge becomes one of reducing the drainage path length. The usual solution is to install an array of vertical drains through the clay with a grid spacing sufficiently small to give the required rate of consolidation. Vertical drains might consist of long flexible porous

7.4 Worked examples

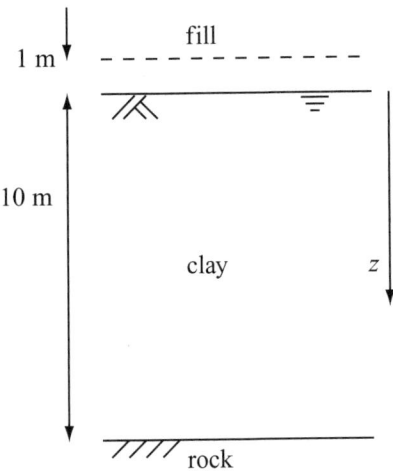

Figure 7.12. Soil conditions for Example 2.

tubes filled with sand, or might be formed from some sort of shaped man-made material which will not close under pressure.

Suppose we installed our drains in a triangular array with a spacing $2d$ (Fig. 7.13). This looks as though it should be a two-dimensional problem but in fact, if the drains are close enough, and since seasonal layering in sedimenting soils tends to lead to bulk horizontal permeabilities which are often much greater than bulk vertical permeabilities (Section 5.6), then the flow of water to dissipate the excess pore pressures takes place predominantly horizontally to the nearest drain – rather equivalent to the steady radial flow to a well that we analysed in Section 5.8. Let us merely note here that the half grid spacing d is roughly equal to the maximum

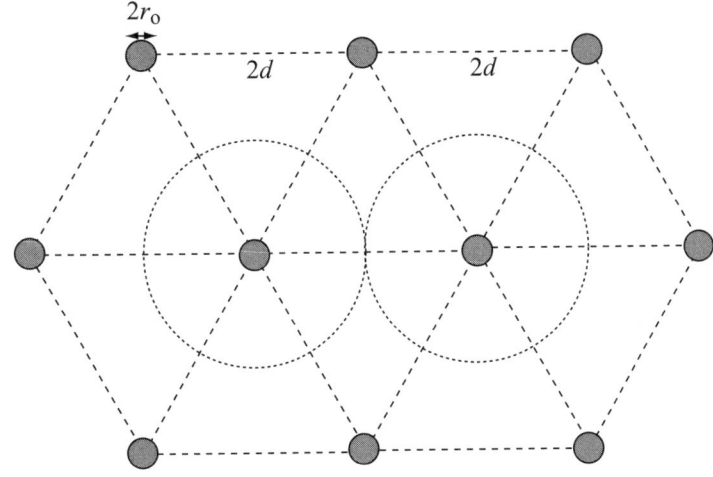

Figure 7.13. Array of vertical drains.

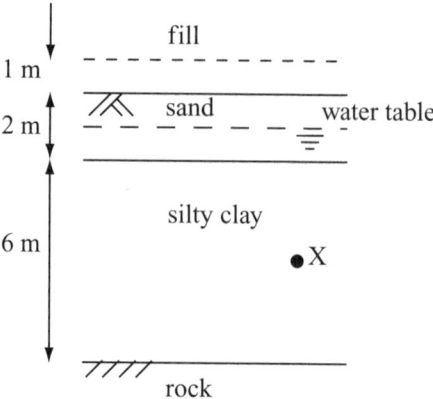

Figure 7.14. Soil conditions for Example 3 (see also Fig. 6.5).

drainage path length.[3] So if we place our drains through our 10 m layer of clay with spacing $2d = 2$ m, then we can expect the time for 90% consolidation settlement to scale by the square of the drainage path length and to be reduced to something of the order of $14.84 \times 1^2/10^2 = 0.15$ years or around 1.8 months, which is likely to be quite acceptable from the point of view of the geotechnical construction.

7.4.3 Example 3

In the example of Section 6.3.3 (Fig. 7.14), we were told that the soils at a certain site consist of 2 m of sand with void ratio $e = 0.6$ overlying 6 m of silty clay with void ratio $e = 1.2$. The silty clay is underlain by impermeable rock. The specific gravity of the particles of both soils is $G_s = 2.7$. The water table is at a depth of 1 m. Above the water table, the sand is dry ($S_r = 0$); below the water table, the sand and silty clay are fully saturated ($S_r = 1$). The site is prepared for construction by the addition of a layer of sandy gravel of thickness 1 m. This material has a void ratio $e = 0.5$ and $G_s = 2.7$. The silty clay has a one-dimensional stiffness $E_o = 1500$ kPa, and we will add that the permeability of the silty clay is $k = 10^{-8}$ m/s. We calculated that the silty clay layer would experience an increase of vertical effective stress of 17.66 kPa and that the eventual settlement of the clay would be 0.071 m. Let us estimate the time that will be taken for 50% and 90% of this settlement to occur.

We need to calculate the coefficient of consolidation, $c_v = kE_o/\gamma_w$, for the clay. From the values given, we have $c_v = 10^{-8} \times 1500/9.81 = 1.53 \times 10^{-6}$ m²/s. For 50% consolidation, we have dimensionless time $\tilde{T} = 0.1792$ from Table 7.2. Our clay layer has thickness $L = 6$ m, since it is underlain by impermeable rock. Hence for 50% consolidation, $t_{50} = 6^2 \times 0.1792/1.53 \times 10^{-6} = 4.22 \times 10^6$ s $= 48.8$ days, and the settlement of the surface of the clay will be 0.035 m. For 90% consolidation, we have dimensionless time $\tilde{T} = 0.7157$ from Table 7.2. Hence for 90% consolidation,

---

[3] The maximum path is evidently a little greater than d and the converging flow to each drain complicates the analysis a little.

7.5 Consolidation: exact analysis ♣

$t_{90} = 6^2 \times 0.7157/1.53 \times 10^{-6} = 16.8 \times 10^6$ s $= 194.9$ days, and the settlement of the surface of the clay will be 0.064 m.

7.4.4 Example 4

At a particular site, there is a clay layer of thickness 5 m sandwiched between layers of sand and gravel, which are able to drain freely. The ground level at the site is to be raised by 2 m by placing a layer of fill of unit weight 21 kN/m^3. To estimate the time that it will take for the clay to settle under this increased stress, a small sample of clay 100 mm in diameter and 20 mm thick is prepared for testing in an oedometer. At an appropriate representative effective stress level, the one-dimensional incremental stiffness is found to be $E_o = 1400$ kPa and it is found that a time of 47 minutes is required for 80% of the settlement to occur under the corresponding stress increment. Calculate the coefficient of consolidation c_v for the clay and the permeability k. How long will it take for 80% of the settlement of the prototype clay layer to occur?

We will use the method of parabolic isochrones. We can use (7.26) to find the value of non-dimensional time $\tilde{T}$ corresponding to a degree of consolidation $\tilde{S} = 0.8$. We need to check that the consolidation process is indeed in the second phase and that the value of $\tilde{T} > 1/12$.

$$\tilde{T} = \frac{1}{12} - \frac{1}{3} \ln \left[\frac{3}{2}(1 - \tilde{S}) \right] \qquad (7.27)$$

and for $\tilde{S} = 0.8$, $\tilde{T} = 0.485$. We know that $\tilde{T} = c_v t/L^2$. For our oedometer sample with drainage from both top and bottom, $L = 0.01$ m and we are told that $t = 47$ min $= 2820$ s. Hence, the coefficient of consolidation $c_v = 1.7 \times 10^{-8}$ m^2/s. The coefficient of consolidation is a function of permeability, stiffness and the unit weight of water: $c_v = kE_o/\gamma_w$ and hence, given $E_o = 1400$ kPa, we can find the permeability $k = 1.2 \times 10^{-10}$ m/s.

For our clay layer *in situ* we need only to know that the time for any chosen degree of consolidation is proportional to the square of the layer thickness, provided that the drainage conditions are equivalent (for example, as here, drainage from top and bottom boundaries in both the oedometer and the prototype). Thus the time for 80% consolidation of the clay layer of thickness $5 = 2 \times 2.5$ m will be $47 \times (2.5/0.01)^2 = 2.94 \times 10^6$ minutes $= 5.59$ years. This would be an unacceptably long time to wait for settlement to be nearing completion, and it would be necessary to take some action to speed up the process – the most obvious route being somehow to reduce the drainage path by installing some artificial network of drains, as we have seen in Section 7.4.2.

7.5 Consolidation: exact analysis ♣

The process that we have just analysed approximately using the assumed parabolic shape of isochrones of excess pore pressure can also be treated exactly. We need to

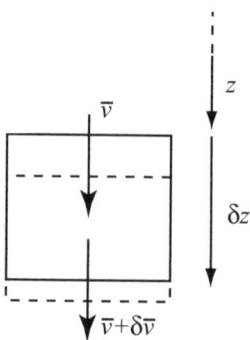

Figure 7.15. One-dimensional consolidation.

develop the governing consolidation equation to describe the coupling between flow and volume change as every element of the soil gradually adjusts to a new effective stress regime. We will restrict ourselves to conditions of one-dimensional flow and deformation.

Our parabolic isochrone analysis equated outflow of water and volume change of the soil resulting from transfer of pore pressure to effective stress for the complete clay layer. The exact analysis applies exactly the same physical reasoning to each infinitesimal element. Flow is governed by Darcy's Law with permeability k, and volume change of the soil is governed by a one-dimensional stiffness E_o.

Our representative element of soil of infinitesimal thickness δz is shown in Fig. 7.15. In general, there will be flow through both top and bottom boundaries of the element according to the gradient of excess pore pressure. Darcy's Law tells us that the velocity of flow $\bar{v}$ is linked with the gradient of total head H or excess pore pressure[4] u' so that, for an element of cross-sectional area A, the volume flow rate into the element at level z will be:

$$q_1 = -A \frac{k}{\gamma_w} \frac{\partial u'}{\partial z} \qquad (7.28)$$

where the negative sign is needed because water flows down the pore pressure gradient, and if $\partial u'/\partial z$ is positive then the flow will be occurring in the $-z$ direction. We have to use the partial differential notation $\partial u'/\partial z$ because we know that the pore pressure is varying with both time and position and we have to distinguish between these two variations.

The volume flow rate into the element at level $z + \delta z$ will be:

$$q_2 = A \frac{k}{\gamma_w} \left[\frac{\partial u'}{\partial z} + \frac{\partial}{\partial z} \left(\frac{\partial u'}{\partial z} \right) \delta z \right] \qquad (7.29)$$

---

[4] Recall the definition of *excess pore pressure* as that transient dissipating pore pressure superimposed on the distribution of pore pressure, associated with steady or zero flow, which we assume to be present before our consolidation event starts and to which the pore pressure will in the long term revert (Section 7.1).

7.5 Consolidation: exact analysis ♣

so that the rate of change of volume V of the element resulting from the flow of water will be:

$$\frac{\partial V}{\partial t} = q_1 + q_2 = A\frac{k}{\gamma_w}\frac{\partial^2 u'}{\partial z^2}\delta z \tag{7.30}$$

The volume of the soil element can only change if the effective stress in the element is changing. From the Principle of Effective stress (Section 2.8), effective stress is the difference between total stress and pore pressure, $\sigma'_z = \sigma_z - u$. We will assume that the total stress is not changing – having placed a load of fill across our site, we are going to sit back and await the effects of consolidation – so the change in effective stress matches the change in pore pressure. There will in general be some *in-situ* equilibrium pore pressure in the ground which is not changing, but it is the excess pore pressure u' which is gradually being dissipated and transferred to effective stress. Thus, we can write:

$$\delta\sigma'_z = -\delta u' \tag{7.31}$$

The changing effective stress results in vertical strain in the element, which is the same as volumetric strain for our one-dimensional system:

$$\delta\varepsilon_z = \frac{\delta\sigma'_z}{E_o} \tag{7.32}$$

The change in volume of the element of cross-section A and thickness δz is:

$$\delta V = -A\delta z\frac{\delta\sigma'_z}{E_o} \tag{7.33}$$

where the negative sign is needed because an increase in effective stress leads to a decrease in volume. The rate of change of volume of the element is:

$$\frac{\partial V}{\partial t} = -\frac{A\delta z}{E_o}\frac{\partial\sigma'_z}{\partial t} = \frac{A\delta z}{E_o}\frac{\partial u'}{\partial t} \tag{7.34}$$

The rate of change in volume of the element resulting from change in effective stress (7.34) must be equal to the rate of change in volume of the element resulting from flow (7.30):

$$\frac{\partial V}{\partial t} = A\frac{k}{\gamma_w}\frac{\partial^2 u'}{\partial z^2}\delta z = \frac{A\delta z}{E_o}\frac{\partial u'}{\partial t} \tag{7.35}$$

or:

$$\frac{kE_o}{\gamma_w}\frac{\partial^2 u'}{\partial z^2} = \frac{\partial u'}{\partial t} \tag{7.36}$$

We should not be surprised to find that the governing soil properties are gathered together in the same way that we found in Section 7.3, defining a coefficient of consolidation (7.15), $c_v = kE_o/\gamma_w$, so that our consolidation equation becomes:

$$c_v\frac{\partial^2 u'}{\partial z^2} = \frac{\partial u'}{\partial t} \tag{7.37}$$

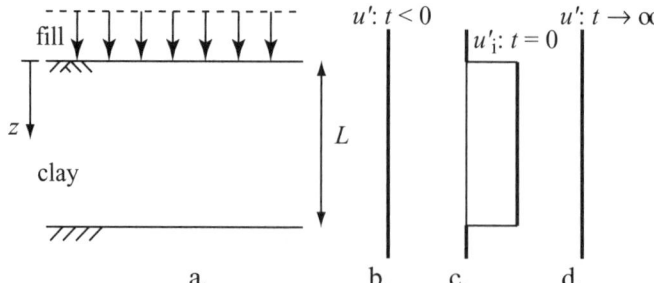

Figure 7.16. One-dimensional consolidation resulting from surcharging of soil layer underlain by impermeable rock. Excess pore pressures at (b) $t < 0$; (c) $t = 0$; (d) $t \to \infty$.

and this is in the form of the standard diffusion equation governing any gradient driven flow.

We can make this equation more universal by replacing our variables u', z and t by non-dimensional variables, as we did for the parabolic isochrones:

$$\tilde{U} = u'/u'_i \tag{7.38}$$

$$\tilde{Z} = z/L \tag{7.39}$$

$$\tilde{T} = \frac{c_v t}{L^2} \tag{7.40}$$

where u'_i is a reference pore pressure – typically the initial uniform excess pore pressure; L is a characteristic length – typically half the thickness of the consolidating layer, if it is drained at both boundaries (Fig. 7.6); and $\tilde{T}$ emerges as a dimensionless time factor. The equation then becomes

$$\frac{\partial^2 \tilde{U}}{\partial \tilde{Z}^2} = \frac{\partial \tilde{U}}{\partial \tilde{T}} \tag{7.41}$$

This equation is based on several assumptions: incompressible pore fluid (so that volume change can only occur because the pore fluid is flowing out of the pores), incompressible soil particles, flow of pore fluid governed by Darcy's Law with a constant coefficient of permeability k, and constant stiffness E_o during the consolidation process. For soft clay soils, with high void ratios and low values of stiffness E_o, the change in geometry incurred during consolidation may be substantial and the assumption of constant E_o and permeability k may also be a little unreasonable.

There are various ways in which the solution of the consolidation equation (7.41) can be developed. Let us consider the simple case of a layer of soil of thickness L over impermeable rock, for which the vertical stress has been increased rapidly over an area of large lateral extent – for example, by placing fill on the ground surface (Fig. 7.16). Since the fill has been placed rapidly at normalised time $\tilde{T} = 0$, there is everywhere an initial excess pore pressure u'_i above the static equilibrium value (Fig. 7.16c) so that, for $\tilde{T} = 0$, $\tilde{U} = 1$ for all $\tilde{Z}$. The overlying soil is assumed to be

7.5 Consolidation: exact analysis ♣

Table 7.4. *Error function* erf(x).

x	erf(x)	x	erf(x)
0	0		
0.1	0.11246	1.1	0.88021
0.2	0.22270	1.2	0.91031
0.3	0.32863	1.3	0.93401
0.4	0.42839	1.4	0.95229
0.5	0.52050	1.5	0.96611
0.6	0.60386	1.6	0.97635
0.7	0.67780	1.7	0.98379
0.8	0.74210	1.8	0.98909
0.9	0.79691	1.9	0.99279
1.0	0.84270	2.0	0.99532

fully drained so that the problem is driven by the reduction of $\tilde{U}$ to zero at $\tilde{Z} = 0$. This is the same case that we analysed using parabolic isochrones in Section 7.3.

7.5.1 Semi-infinite layer

We saw with the parabolic isochrone approximate analysis that there is initially a propagation process as the sensation of the reduction of pore pressure at the surface progressively spreads into the body of the clay. The problem is the same as that of suddenly changing the temperature at one end of a long conducting bar, for which Carslaw and Jaeger[5] show that the solution can be written:

$$\tilde{U} = \mathrm{erf}\left(\frac{\tilde{Z}}{2\sqrt{\tilde{T}}}\right) \quad (7.42)$$

where erf(x) is the error function:

$$\mathrm{erf}(x) = \frac{2}{\sqrt{\pi}} \int_0^x e^{-y^2}\, dy \quad (7.43)$$

for which some values are given in Table 7.4 and plotted in Fig. 7.17a. It may seem mysterious to produce this error function solution apparently out of the air, but substitution will confirm that (7.42) is indeed a solution of (7.41) which satisfies the boundary conditions of the semi-infinite layer ($\tilde{U} = 0$ for $\tilde{Z} = 0$ for all $\tilde{T}$; $\tilde{U} \to 1$ as $\tilde{Z} \to \infty$ for all $\tilde{T}$). This substitution requires the rule for the differentiation of a definite integral:

$$\frac{\partial}{\partial x}\int_{a(x)}^{b(x)} f(x, y)\, dy = \int_{a(x)}^{b(x)} \frac{\partial f}{\partial x}\, dy + f(x, b(x))\frac{\partial b}{\partial x} - f(x, a(x))\frac{\partial a}{\partial x} \quad (7.44)$$

The error function is within 1% of unity for values of the argument greater than about 2. Therefore, from (7.42) we can deduce that the normalised depth $\tilde{Z}_p$ to

---

[5] Carslaw, H.S. & Jaeger, J.C. (1959) *Conduction of heat in solids*, (2nd ed.) Clarendon Press, Oxford.

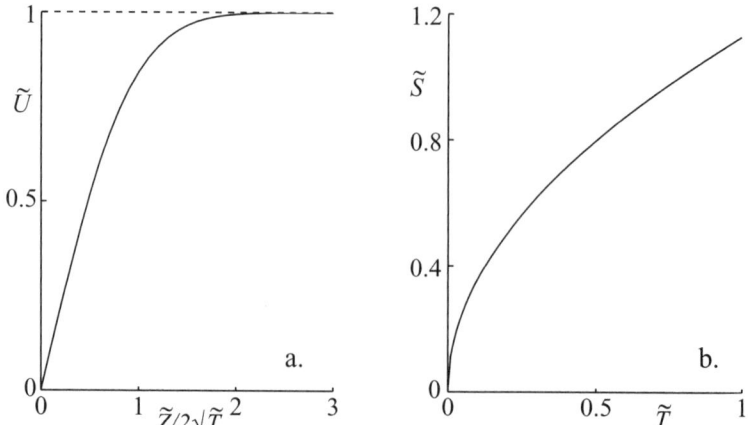

Figure 7.17. One-dimensional consolidation, semi-infinite layer: (a) dimensionless variation of pore pressure with depth; (b) dimensionless variation of degree of consolidation with time.

which the consolidation front has penetrated at any time $\tilde{T}$ is given by:

$$\tilde{Z}_p \approx 4\sqrt{\tilde{T}} \qquad (7.45)$$

which can be compared with our estimate using parabolic isochrones (7.23):

$$\frac{\ell}{L} = \sqrt{12\tilde{T}} \qquad (7.46)$$

The error function is steeper than the parabolic isochrone, and the consolidation front has to penetrate further to balance the volume change resulting from the outflow.

While it is interesting to know how far the consolidation process has penetrated, it is practically more useful to know how the settlement develops at the surface of the consolidating soil. A reference settlement is required: the settlement of a finite layer of characteristic thickness L ($\tilde{Z}=1$) in which the pore pressure falls from u'_i ($\tilde{U}=1$) to zero and the effective stress increases by a corresponding amount. The reference settlement is then:

$$\varsigma_\infty = \frac{u'_i L}{E_o} \qquad (7.47)$$

and we can define a dimensionless settlement (recall (7.24)):

$$\tilde{S} = \frac{\varsigma}{\varsigma_\infty} = \frac{E_o \varsigma}{u'_i L} \qquad (7.48)$$

Settlement of the layer occurs because the soil is becoming more compressed as water is squeezed out at the surface of the layer. The rate at which water leaves the soil is governed, through Darcy's Law, by the pore pressure gradient at the surface of the layer. Combining (7.2) and (7.4):

$$\frac{d\varsigma}{dt} = \frac{k}{\gamma_w} \left(\frac{\partial u'}{\partial z}\right)_{z=0} \qquad (7.49)$$

7.5 Consolidation: exact analysis ♣

which we can convert to a equation in terms of dimensionless quantities:

$$\frac{d\tilde{S}}{d\tilde{T}} = \frac{E_o}{u'_i L} \frac{L^2}{c_v} \frac{d\varsigma}{dt} = \frac{E_o}{u'_i L} \frac{L^2}{c_v} \frac{k}{\gamma_w} \left(\frac{\partial u'}{\partial z}\right)_{z=0} = \frac{L}{u'_i} \left(\frac{\partial u'}{\partial z}\right)_{z=0} = \left(\frac{\partial \tilde{U}}{\partial \tilde{Z}}\right)_{\tilde{Z}=0} \quad (7.50)$$

Differentiating (7.42) (noting (7.44)):

$$\left(\frac{\partial \tilde{U}}{\partial \tilde{Z}}\right)_{\tilde{Z}=0} = \frac{1}{\sqrt{\pi \tilde{T}}} \quad (7.51)$$

and hence:

$$\tilde{S} = \sqrt{\frac{4\tilde{T}}{\pi}} \quad (7.52)$$

which is plotted in Fig. 7.17b. The degree of consolidation varies with the square root of time (compare $\tilde{S} = \sqrt{4\tilde{T}/3}$ for parabolic isochrones (7.24)).[6]

This first analysis will be valid also for a finite clay layer, of normalised thickness $\tilde{Z} = 1$, provided that the distance that the consolidation front has penetrated is less than the thickness of the layer or, approximately, from (7.45), $\tilde{T} < 1/16$ or $t < L^2/16c_v$ – which can be compared with our parabolic isochrone estimate of the end of this stage of consolidation occurring at $\tilde{T} = 1/12$.

7.5.2 Finite layer

If we know from the beginning that we are dealing with a finite layer, then we can obtain a single analytical solution to the governing equation (7.41). Let us repeat the definition of the problem. A finite layer of compressible soil of thickness L is underlain by impermeable rock (Fig. 7.16). The pore pressure throughout the layer is initially equal to u'_i ($\tilde{U} = 1$ at $t = \tilde{T} = 0$). The governing partial differential equation can be solved using the standard technique of separation of variables.

We assume that the solution will be of the form $\tilde{U} = \mathcal{Z}(\tilde{Z})\mathcal{T}(\tilde{T})$, where $\mathcal{Z}(\tilde{Z})$ and $\mathcal{T}(\tilde{T})$ are functions only of $\tilde{Z}$ and $\tilde{T}$, respectively. Then:

$$\frac{\partial^2 \tilde{U}}{\partial \tilde{Z}^2} = \mathcal{T}\frac{d^2 \mathcal{Z}}{d\tilde{Z}^2} \quad (7.53)$$

and:

$$\frac{\partial \tilde{U}}{\partial \tilde{T}} = \mathcal{Z}\frac{d\mathcal{T}}{d\tilde{T}} \quad (7.54)$$

so that:

$$\mathcal{T}\frac{d^2 \mathcal{Z}}{d\tilde{Z}^2} = \mathcal{Z}\frac{d\mathcal{T}}{d\tilde{T}} \quad (7.55)$$

---

[6] Note that $\tilde{U}$ is a local variable describing the normalised excess pore pressure, whereas $\tilde{S}$ is a system variable describing the overall settlement of a particular consolidating system.

Since this must apply for all $\tilde{Z}$ and all $\tilde{T}$, we can rearrange and deduce that solutions must satisfy:

$$\frac{1}{\mathcal{Z}}\frac{d^2\mathcal{Z}}{d\tilde{Z}^2} = \frac{1}{\mathcal{T}}\frac{d\mathcal{T}}{d\tilde{T}} = -\omega^2 \tag{7.56}$$

where ω takes values necessary to satisfy the boundary conditions of our problem. We now have two ordinary differential equations to solve.

The equation:

$$\frac{d^2\mathcal{Z}}{d\tilde{Z}^2} + \omega^2 \mathcal{Z} = 0 \tag{7.57}$$

is the equation of simple harmonic motion with general solution:

$$\mathcal{Z} = J_1 \sin \omega \tilde{Z} + J_2 \cos \omega \tilde{Z} \tag{7.58}$$

But we know that the pore pressure is zero at $\tilde{Z} = 0$ at all times, so $J_2 = 0$. We also know that there is no flow through the impermeable boundary at $\tilde{Z} = 1$ ($z = L$), which requires that the solution must be built up from terms which are formed only of odd harmonics:

$$\omega = (2m+1)\frac{\pi}{2} \tag{7.59}$$

where m can take any integer value $m \geq 0$.

The other part of the equation describes the time dependency of the pore pressure change:

$$\frac{d\mathcal{T}}{d\tilde{T}} + \omega^2 \mathcal{T} = 0 \tag{7.60}$$

with solution

$$\mathcal{T} = J_m \exp\left[-\omega^2 \tilde{T}\right] = J_m \exp\left[-\pi^2(2m+1)^2 \tilde{T}/4\right] \tag{7.61}$$

and the complete solution is the sum of terms of the form:

$$\tilde{U} = \sum_{m=0}^{\infty}\left\{J_m \exp\left[-\pi^2(2m+1)^2 \tilde{T}/4\right] \sin\left[(2m+1)\frac{\pi}{2}\tilde{Z}\right]\right\} \tag{7.62}$$

We have one other boundary condition to apply. At time $t = \tilde{T} = 0$, the non-dimensional pore pressure is constant throughout the layer $\tilde{U} = \tilde{U}_o = 1$ for $0 < \tilde{Z} < 1$. Thus:

$$\tilde{U}_o = 1 = \sum_{m=0}^{\infty}\left\{J_m \sin\left[(2m+1)\frac{\pi}{2}\tilde{Z}\right]\right\} \tag{7.63}$$

We can find the values of the coefficients J_m by multiplying both sides of (7.63) by $\sin\left[(2n+1)\frac{\pi}{2}\tilde{Z}\right]$ and integrating over the range $0 < \tilde{Z} < 1$. The right-hand side

7.5 Consolidation: exact analysis ♣

consists of terms of the form:

$$J_m \int_0^1 \sin\left[(2m+1)\frac{\pi}{2}\tilde{Z}\right]\sin\left[(2n+1)\frac{\pi}{2}\tilde{Z}\right]d\tilde{Z}$$

$$= J_m \int_0^1 \frac{1}{2}\left\{\cos\left[(2m-2n)\frac{\pi}{2}\tilde{Z}\right] - \cos\left[(2m+2n+2)\frac{\pi}{2}\tilde{Z}\right]\right\}d\tilde{Z}$$

$$= J_m \frac{1}{2}\left\{\frac{2}{\pi(2m-2n)}\sin\left[(2m-2n)\frac{\pi}{2}\tilde{Z}\right]\right. \tag{7.64}$$

$$\left. - \frac{2}{\pi(2m+2n+2)}\sin\left[(2m+2n+2)\frac{\pi}{2}\tilde{Z}\right]\right\}$$

$$= 0 \quad \text{for} \quad m \neq n$$

since m and n are integers. For $m = n$, the only remaining term on the right-hand side of (7.64) becomes:

$$J_m \int_0^1 \sin^2\left[(2m+1)\frac{\pi}{2}\tilde{Z}\right]d\tilde{Z} = J_m \int_0^1 \frac{1}{2}\left\{1 + \cos\left[(2m+1)\pi\tilde{Z}\right]\right\}d\tilde{Z} = \frac{J_m}{2} \tag{7.65}$$

The left-hand side is:

$$\int_0^1 \sin\left[(2m+1)\frac{\pi}{2}\tilde{Z}\right]d\tilde{Z} = \frac{2}{(2m+1)\pi}\left[-\cos\left\{(2m+1)\frac{\pi}{2}\tilde{Z}\right\}\right]_0^1 = \frac{2}{(2m+1)\pi} \tag{7.66}$$

Hence:

$$J_m = \frac{4}{(2m+1)\pi} \tag{7.67}$$

and the complete Fourier series solution of the consolidation equation is:

$$\tilde{U} = \frac{4}{\pi}\sum_{m=0}^{\infty}\left\{\frac{1}{2m+1}\exp\left[-\pi^2(2m+1)^2\frac{\tilde{T}}{4}\right]\sin\left[(2m+1)\frac{\pi}{2}\tilde{Z}\right]\right\} \tag{7.68}$$

where m takes integral values from 0 to ∞. The degree of consolidation $\tilde{S}$ at time $\tilde{T}$ is found from the integration of $\tilde{U}$ over the range $0 < \tilde{Z} < 1$:

$$\tilde{S} = 1 - \frac{8}{\pi^2}\sum_{m=0}^{\infty}\left\{\frac{1}{(2m+1)^2}\exp\left[-\pi^2(2m+1)^2\frac{\tilde{T}}{4}\right]\right\} \tag{7.69}$$

This Fourier series solution considers the entire clay layer throughout the analysis and is not at all concerned with the propagation of disturbances into the layer from a boundary. It can be used to generate a family of isochrones showing the spatial variation of pore pressure at different times (Fig. 7.18a). The Fourier series solution struggles to match the initial condition of uniform pore pressure through the summation of harmonic functions: a very large number of terms from the infinite series (7.68, 7.69) has to be included to begin to produce an accurate representation of the early stages of the analysis and to match the physical constraints. Figure 7.19 compares the effect of taking the first 10 terms or the first 500 terms of the Fourier

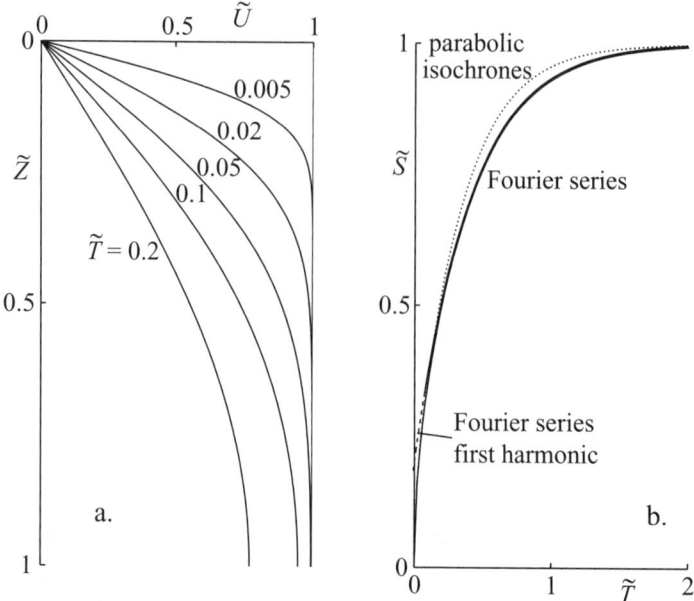

Figure 7.18. One-dimensional consolidation. (a) Isochrones of pore pressure from Fourier series solution; (b) degree of consolidation from Fourier series and parabolic isochrone solution.

series to try to match the initial pore pressure isochrone, which should have $\tilde{U} = 1$ for all $\tilde{Z}$. Even with 500 terms, the summation is inexact for small values of $\tilde{Z}$.

As time goes by, however, the exponential term decays rapidly because of the multiplier $(2m+1)^2$ and the first harmonic becomes dominant (Fig. 7.18a), and it is

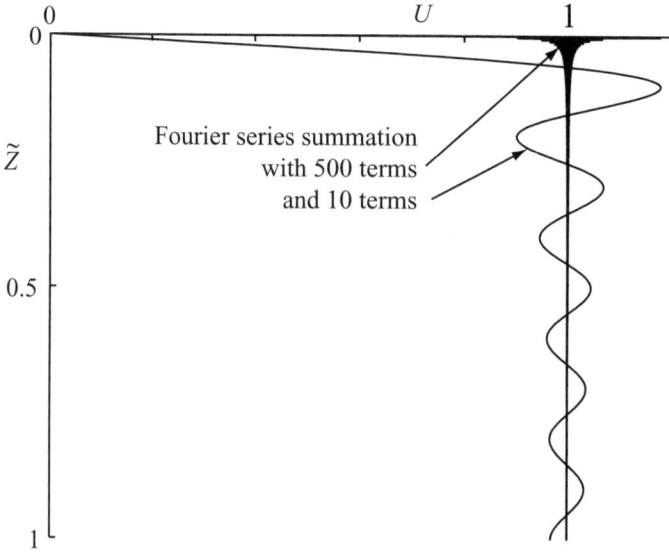

Figure 7.19. Fourier series solution for isochrone of pore pressure for $\tilde{T} = 0$; summation of 10 terms and 500 terms.

sufficient to consider just the first term of the Fourier series:

$$\tilde{U} = \frac{4}{\pi} \exp\left[-\pi^2 \frac{\tilde{T}}{4}\right] \sin \frac{\pi \tilde{Z}}{2} \tag{7.70}$$

and:

$$\tilde{S} = 1 - \frac{8}{\pi^2} \exp\left[-\pi^2 \frac{\tilde{T}}{4}\right] \tag{7.71}$$

This relationship is also plotted in Fig. 7.18b together with the exact result (7.69) and with the result obtained using the parabolic isochrone approximation. The first harmonic approximation is obviously in error for low values of $\tilde{T}$ (it suggests that $\tilde{S} = 0.189$ for $\tilde{T} = 0$) but is extremely close for $\tilde{T}$ greater than about 0.1.

The exact Fourier series solution may be cumbersome to use but there are other solutions which are quite accurate within certain ranges of the problem and which are much simpler to work with. For small times, the theory for an infinite layer (7.42, 7.52) is appropriate, describing the propagation of a consolidation front into the soil from the drainage boundary; for large times, the finite layer theory with a single harmonic to describe the spatial variation of pore pressure is sufficient (7.70, 7.71). The parabolic isochrone approach gives a good approximation to both stages of the analysis.

7.6 Summary

Here is a concise list of the key messages from this chapter, which are also encapsulated in the mind map (Fig. 7.20).

1. Consolidation describes the time-dependent transient process of soil deformation as total stress changes are transferred from pore pressure to effective stress.
2. Soil deformation requires volume change which, in saturated soils, requires flow of pore fluid (water) which will be slow in low permeability clayey soils. Flow of pore water is governed by Darcy's Law and depends on the gradient of total head or excess (non-equilibrium) pore pressure. The process of consolidation is encapsulated in a partial differential diffusion equation linking spatial and temporal variations of pore pressure.
3. The diffusion equation appears in many different physical systems where the key phenomenon is one of gradient-driven flow.
4. The solution of the equation can be presented in terms of isochrones which show the spatial variation of pore pressure at chosen times.
5. An approximate solution can be obtained assuming that the isochrones have a parabolic shape at all times: the two stages of the process of consolidation can then be described by two first order ordinary differential equations.
6. There are various different analytical techniques that can be used to study the progress of consolidation. The first phase of consolidation involves the propagation into the clay of the consolidation front – the sensation of the presence of a distant drainage boundary.

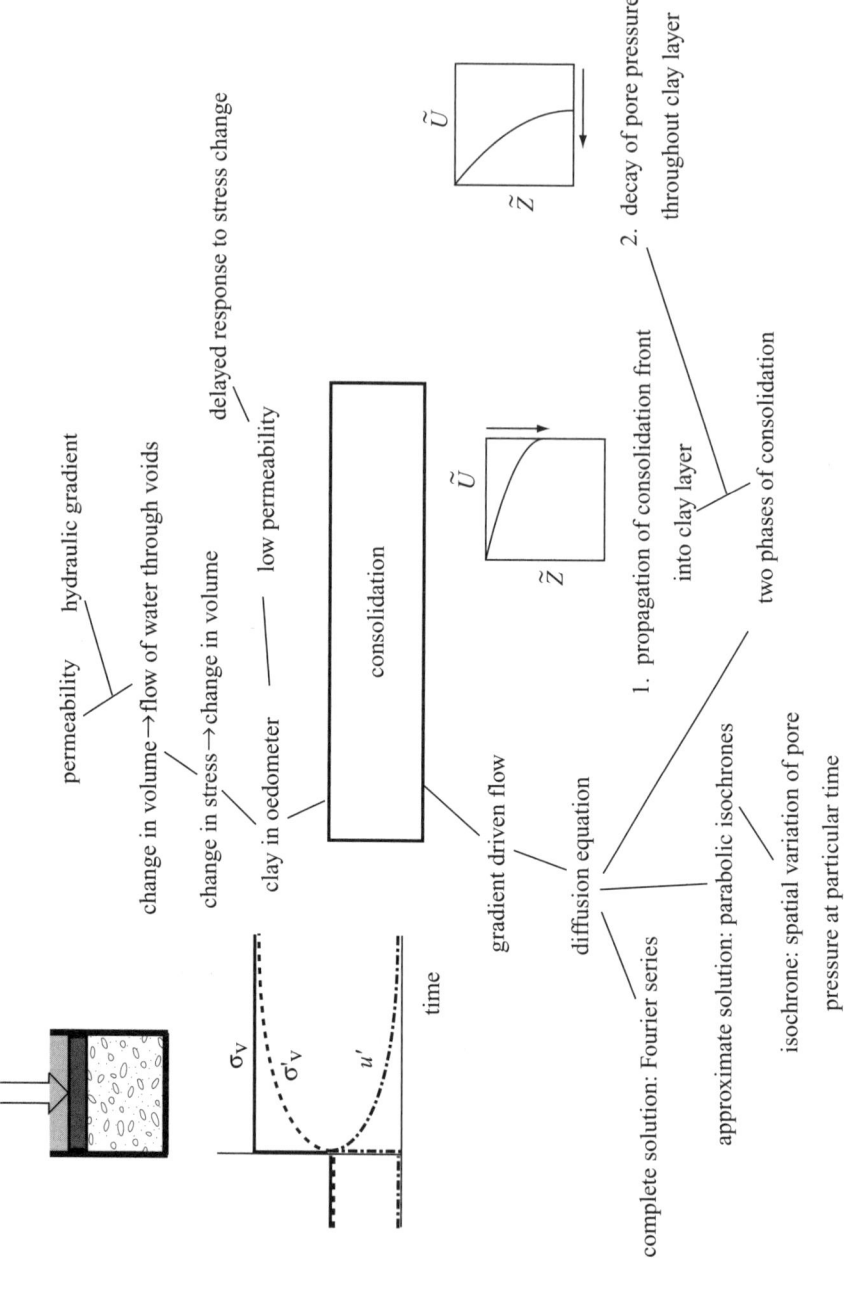

Figure 7.20. Mind map: consolidation.

7.7 Exercises: Consolidation

7. When the complete clay layer is participating in the consolidation process, then the isochrones take on a simple sinusoidal mode shape.

7.7 Exercises: Consolidation

Make use of the simplified parabolic isochrones analysis to solve Problems 1–4.

1. The sample of clay in Section 6.5, Questions 1/2 has a thickness of 20 mm. Drainage is able to occur from both the top and bottom boundaries of the sample. The permeability of the clay is 10^{-9} m/s. What is the value of the coefficient of consolidation c_v? How long after the change in applied total stress does the clay at the centre of the sample become aware of the consolidation process? How long after the change in applied total stress has the settlement of the clay reached 80% of its final magnitude?

2. The sample of clay in Section 6.5, Questions 1/2 is representative of a clay layer with a thickness of 3 m contained between layers of sand and gravel. The site is being prepared by rapidly placing a layer of fill at the ground surface. How long after the placement of this fill will the settlement of the clay reach 80% of its final magnitude? Does this time depend on the magnitude of the stress change generated by the placement of the fill?

3. The sample of clay in Section 6.5, Questions 4/5 is found to reach 50% of its final consolidation settlement in 95 minutes. It is representative of a clay layer of thickness 2.4 m. How long will it take for 90% settlement of this clay layer to occur?

4. Oedometer tests on a sample from the bed of clay described in Section 6.5, Question 7 show that, for an appropriate stress increment, the coefficient of permeability is 10^{-9} m/s. How long will it take for the effect of the removal of the top 1 m of sand to be felt as a change of effective stress at the middle of the clay?

5. Any mathematical shape that satisfies the boundary conditions could be assumed for the typical isochrone of excess pore pressure in Fig. 7.8. In this chapter we have assumed a parabolic form. Repeat the analysis of the two stages of consolidation for a layer of thickness L with drainage only from the upper boundary using a sinusoidal function so that in the first stage the typical isochrone is $u' = u'_i \sin \pi z/2\ell$ and in the second stage the typical isochrone is $u' = u'_L \sin \pi z/2L$. Find expressions for the variation of degree of consolidation $\tilde{S} = \varsigma/\varsigma_\infty$ with dimensionless time $\tilde{T} = c_v t/L^2$ during the two stages.

6. Results of one increment of loading (from 200 to 400 kPa) in an oedometer test on reconstituted Gault clay are given in Table 7.5. At the start of the increment the height of the sample was 15.56 mm. The sample was able to drain from top and bottom.
Using a procedure similar to that described in Section 7.4.1, use the exact analysis encapsulated in (7.52) and (7.71) to estimate the one-dimensional stiffness, coefficient of consolidation, and permeability of the clay. Estimate the

Table 7.5. *Data from oedometer test on reconstituted Gault clay.*

Time minutes	Settlement mm	Time minutes	Settlement mm
0	0	25	0.6934
0.25	0.0635	30.25	0.7468
1	0.1372	36	0.7899
2.25	0.2032	42.25	0.8306
4	0.2718	49	0.8634
6.25	0.3480	56.25	0.8909
9	0.4242	64	0.9130
12.25	0.4978	72.25	0.9273
16	0.5664	81	0.9389
20.25	0.6325	90.25	0.9504

time taken for 90% consolidation of a layer of clay 5 m thick, drained top and bottom.

7. Repeat Question 6 using the approximate analysis with sinusoidal shape functions proposed in Question 5.

8 Strength

8.1 Introduction

In the context of our chosen one-dimensional approach to the mechanics of soils, we are somewhat limited in what we can say about the strength of soils but there are some ideas which can usefully be presented. Stiffness is concerned with the deformations of geotechnical systems – the *serviceability limit states* under operational or working loads. Strength is concerned with the collapse of geotechnical systems – the *ultimate limit states* for which failure of the geotechnical system will occur. Classically, it has always been easier to make statements about ultimate collapse conditions than about deformations, and geotechnical design often proceeds by starting with a collapse calculation and then factoring down the loading sufficiently that, from experience, the resulting reduced load would not be expected to produce excessive displacements. This is always a rather uncertain route by which to control those displacements, especially if the nature of the problem under consideration is more than somewhat different from those previously experienced – a proper understanding of stiffness is really more satisfactory. However, it does emphasise the traditional importance of understanding the strength of soils and the modes of failure of geotechnical systems.

8.2 Failure mechanisms

Figure 8.1 shows a schematic picture of a pile foundation. A *pile* is a long slender stiff structural member which is used to transfer loads from some surface structure through more or less soft soils to a certain depth in the ground (Fig. 1.9). The surface load P is shared with the soil over the length of the pile by the generation of shear stresses τ – the so-called *shaft resistance* – and by some *end bearing* stress σ_b at the base of the pile. The relative proportions of the load taken through shaft resistance and end bearing will depend on the ground conditions but we can imagine that, if the load P is increased until the pile fails, then, at failure, the pile is being pushed uncontrollably into the ground, and there will be some limiting strength of the soil τ_f being mobilised down the length of the pile at the interface between the pile (steel or concrete or timber) and the soil.

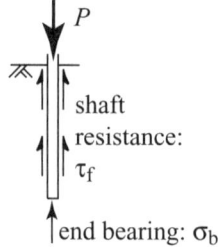

Figure 8.1. Pile foundation.

There are various circumstances in which a slope will fail by sliding on a clear slip surface as suggested in Figs 8.2[1], 8.3. If this happens, there is a block of soil which is moving relative to the unfailing soil on some slip surface, and we expect that there will be some shear stress on this slip surface resisting the movement. The slip surface may be somewhat circular (Fig. 8.3a) or may be constrained by the presence of some strong layer to be distinctly non-circular (Fig. 8.3b) but in either case the failure is controlled by the shear strength τ_f that can be mobilised at different points along the slip surface.

Where the soil near the surface of the ground has adequate strength, a shallow foundation (rather than a pile, which would be thought of as a deep foundation) may be a preferred method of conveying the loads from a structure to the ground. Figure 8.4 hints at the possibility of a failure occurring by rotation of the

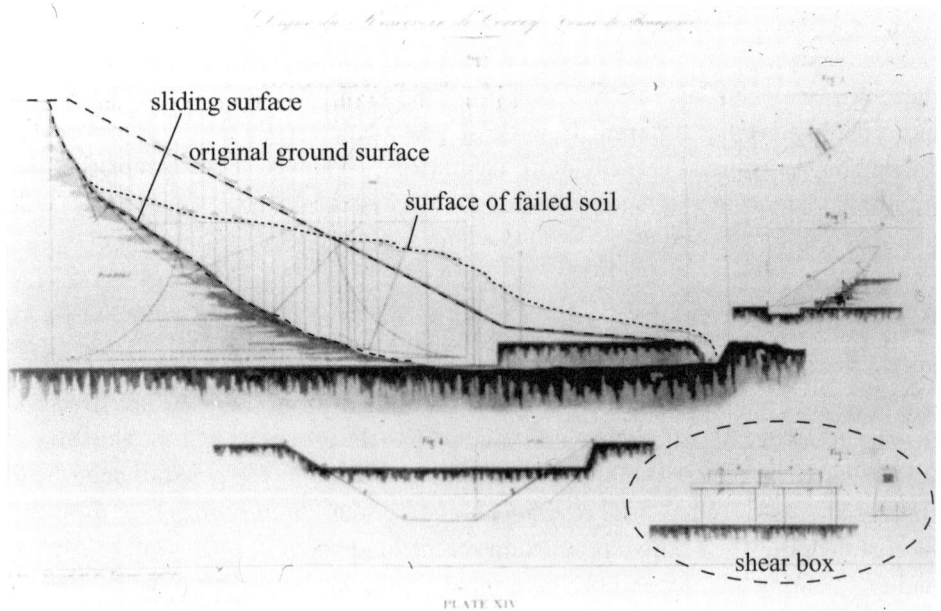

Figure 8.2. Sketches by Collin (1846)[1] of slope failure in stiff clay.

[1] Collin, A. (1846) *Recherches expérimentales sur les glissements spontanés des terrains argileux*. Paris, Carilian-Gœurley et Dalmont.

8.3 Shear box and strength of soils

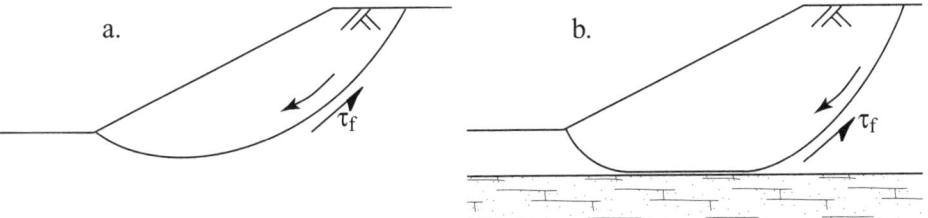

Figure 8.3. Slope failure, block of soil sliding with well-defined slip surface: (a) circular mechanism; (b) non-circular mechanism.

foundation, once again mobilising some limiting shear stress τ_f along the sliding surface separating moving soil from the soil which is left behind.

What these examples all have in common is the suggestion that failure, when it occurs – by increasing the load on the deep or shallow foundation or by increasing the height or steepening the angle of a slope – may often occur by the formation of a mechanism of failure within the soil which involves sliding along a defined surface in the soil.[2] With that observation of actual failure mechanisms in mind, one of the early pieces of equipment developed for determining the strength of soils in the laboratory (by Collin in the 1840s) was the shear box (Figs 8.5, 8.6). This piece of equipment remains in widespread use today and has excellent pedagogic value for the insights it can give into the behaviour of soils.

8.3 Shear box and strength of soils

The shear box is a laboratory test device which is used to force a failure surface to form within a soil sample contained within a split box (Fig. 8.6a). Typical shear boxes might have cross-sections 60×60 mm, or 100×100 mm or 300×300 mm (or larger) depending on the largest sizes of the particles in the soil being tested. The upper and lower parts of the box are initially aligned and filled with the soil. The soil is then loaded through a platen with a force P which generates a vertical stress $\sigma_z = P/A$, where A is the cross-sectional area of the failure plane. There may be a

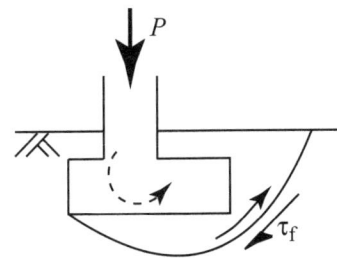

Figure 8.4. Failure of shallow foundation.

---

[2] However, this is not inevitable and sometimes geotechnical failures will occur by some more diffuse mode of deformation for which the model of soil behaviour developed in Section 8.11 is appropriate.

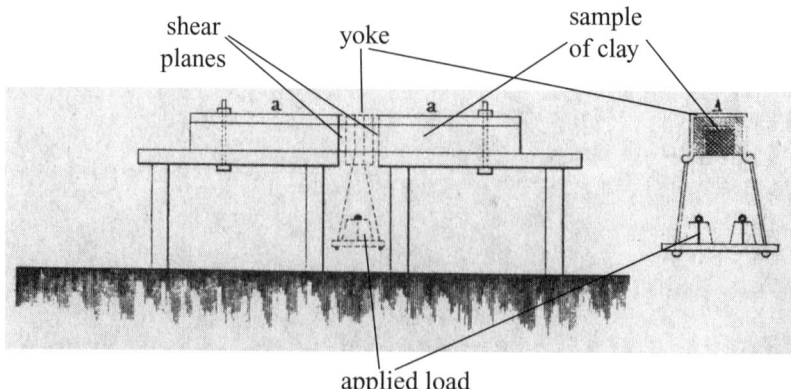

Figure 8.5. Early "shear box" used by Collin to discover the undrained shear strength of clay (drawn from diagram in bottom right corner of Fig. 8.2).

pore pressure u around the failure region (although the construction of the shear box makes it difficult to sustain any pore pressure in a moderately permeable soil) so that the effective normal stress on the failure plane is $\sigma'_z = \sigma_z - u$. The upper and lower parts of the box are forced apart by applying a relative displacement x and the resulting shear force Q resisting the failure of the soil is measured. The way in which the apparatus is constructed means that the area of contact between the two halves of the box diminishes as the sample is sheared, and this fact should be taken into account as we calculate the vertical stress. The horizontal load Q is applied in line with the failure plane and generates a shear stress $\tau = Q/A$ (Fig. 8.6b). We assume that the normal and shear stresses calculated in this way are representative of the average stresses on the failure plane (in practice, the distribution of stresses will be far from uniform). It is also a good discipline to measure the downward vertical movement z of the upper half of the box as the shear displacement x is increased.

A typical set of test results from shear box tests on dense and loose sands is shown in Fig. 8.7.[3] Figure 8.7a shows the development of the shear force Q

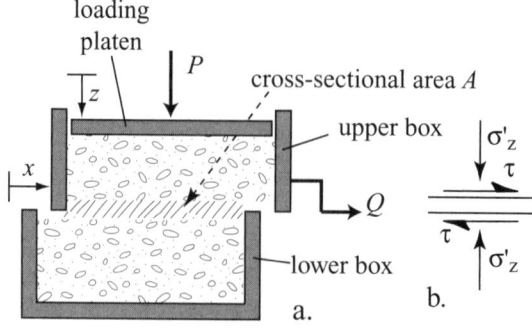

Figure 8.6. Shear box.

[3] Taylor, D.W. (1948) *Fundamentals of soil mechanics.* John Wiley, New York.

8.4 Strength model

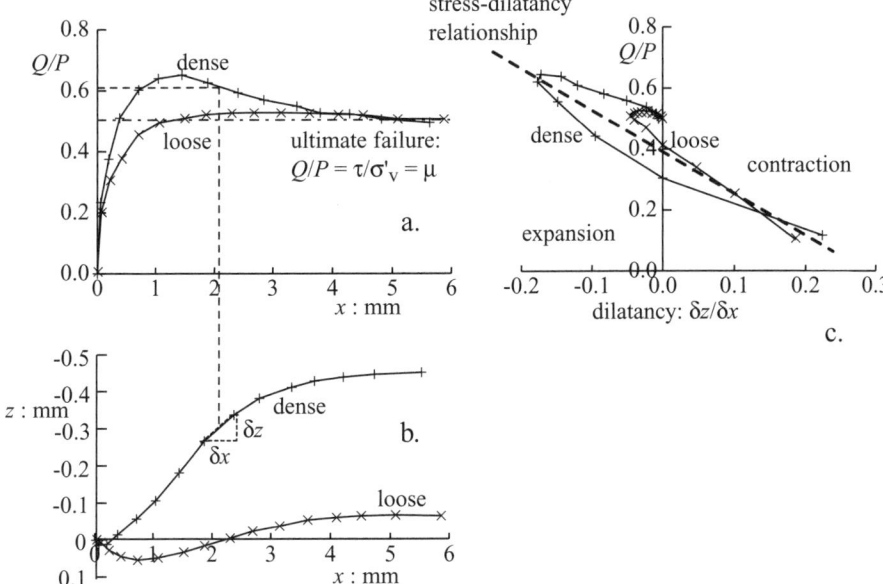

Figure 8.7. (a) Shear load:displacement response and (b) volume changes in direct shear test on Ottawa sand; (c) stress-dilatancy correlation (see Section 8.11.4) (data from Taylor[3]).

normalised with the normal load P, $Q/P = \tau/\sigma'_z$, with shear displacement x. The sand is dry so that the pore pressure $u = 0$. The shear force has a clear peak and also, as the shearing continues, appears to approach an ultimate asymptotic value. There are various ways in which we might define *strength*. From the point of view of choosing a strength for the soil that might be used for the purposes of design, the peak value of shear stress (converted from the maximum shear force) appears to have some attraction but we observe that it presents an ephemeral benefit: it is not a strength that can be sustained or relied upon as shearing continues. The asymptotic value of shear stress seems to be much more reliable and it is that *ultimate failure* value that we will build into our strength model and for which we will use the designation τ_f.

8.4 Strength model

We will describe here a simple strength model for soils which builds on intuition to make some physically reasonable assumptions. The shear box (Fig. 8.6) has two degrees of freedom: the normal load P, which pushes down on the top platen of the box and holds the sample together, with associated displacement z; and the shear force Q, which causes the sample to divide into two halves, with associated displacement x. We would expect that, the more we push down on the top of the sample, the more difficult it will be to cause this shear failure to occur. Let us suppose that the shear strength is directly proportional to the normal force, or in terms of stresses (because it is more useful to work in terms of areal intensity of forces) (Fig. 8.8a):

$$\tau_f = \mu \sigma'_z \tag{8.1}$$

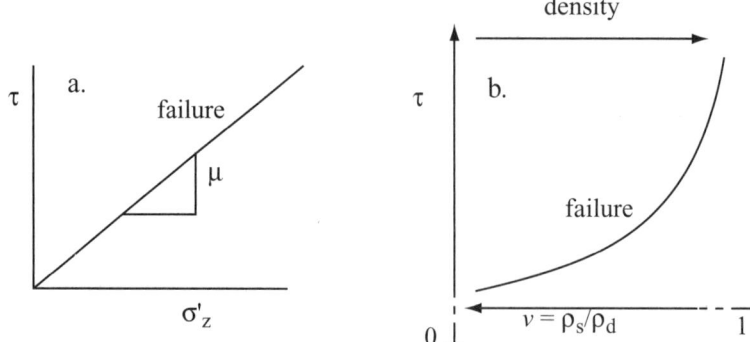

Figure 8.8. Strength model: (a) frictional relationship between shear stress and effective normal stress; (b) dependence of strength on soil density or volumetric packing.

We expect the shear strength τ_f to be proportional to the vertical effective stress $\sigma'_z = P/A - u$ because we know that it is effective stresses that control all aspects of the stiffness and strength of soils. The constant of proportionality μ indicates that we are proposing a frictional model: we could write $\mu = \tan\phi$, where ϕ is an angle of friction or shearing resistance of the soil. We will continue to use σ'_z to indicate the normal effective stress acting on our failure surface but need to recognise that, in general, the failure surface may be anything but horizontal (Figs 8.1, 8.2, 8.3).

The second aspect of the model is to propose, again using our intuition, that the strength of soils will depend on their density or volumetric packing (Fig. 8.8b). Volumetric packing is indicated by the specific volume v, which is the ratio of soil mineral density ρ_s to soil dry density ρ_d (Section 3.3.3) and thus proportional to the reciprocal of the density of the soil. Specific volume falls towards unity as the density of the soil increases towards the density of the soil mineral and the voids surrounding the particles become smaller and smaller. The relationship between strength and density is probably non-linear but we can draw some quite important conclusions even without actually assuming a form for the curve in Fig. 8.8b.

The pair of diagrams in Fig. 8.8 suggests that we could define a unique three-dimensional curve describing the strength in terms of the limiting shear stress τ_f, effective normal stress σ'_z and specific volume v, and could propose that in any shearing test the soil would be trying to reach a strength point on this curve. What are the consequences of such a proposition? And is it a reasonable proposition? We can return to the diagram in Fig. 8.7b which shows the record of vertical movement z in the shear box tests and introduce the concept of dilatancy.

8.5 Dilatancy

We have seen that the chief characteristic which distinguishes most soils from other engineering materials (such as metals and plastics) is the high proportion of the volume of the material which is made up of void filled with a single or multi-phase fluid (for example, air or water, Section 3.3). For a typical medium dense sand, about a third of the overall volume is void; for a normally consolidated clay, voids might

8.5 Dilatancy

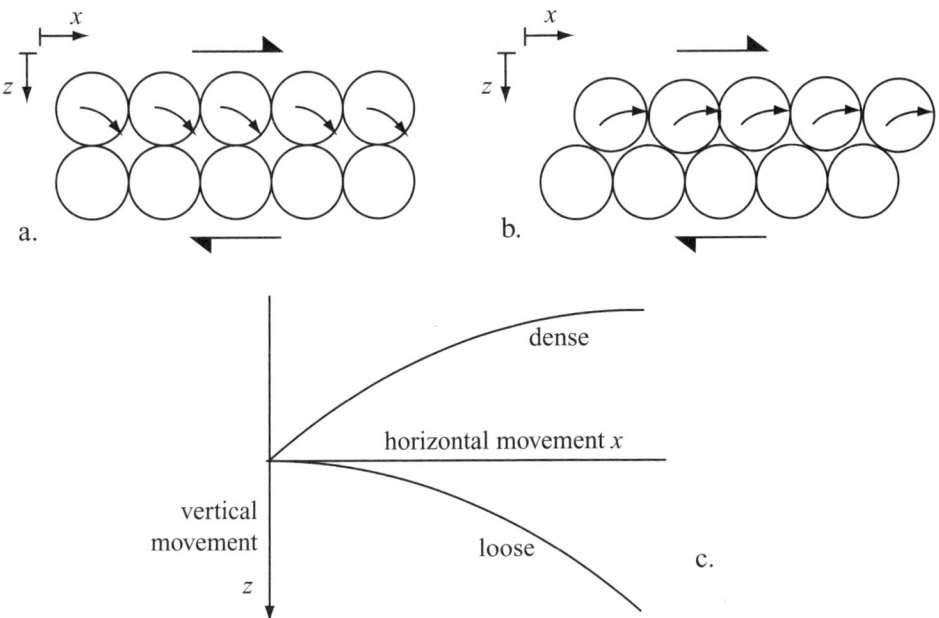

Figure 8.9. (a) Shearing of loosely packed layers of circular discs; (b) shearing of densely packed layers of circular discs; (c) volume change (=vertical movement) in shearing of loosely and densely packed layers of circular discs.

contribute half of the volume. We can understand that one-dimensional compression results in volume changes and increase in density when we increase the vertical effective stress (Section 4.5), but we can also see that when a sand is sheared in the shear box (Fig. 8.6) there is vertical movement of the top half of the box indicating variation of density of the sand as it is being sheared (Fig. 8.7b). A simple, imaginary test on a two-dimensional analogue of a soil can be used to demonstrate why these changes in volume occur when all we are trying to do is to move one block of soil horizontally relative to another block of the soil.

Figure 8.9a shows a loose packing of circular particles in two layers. As the upper layer is moved sideways relative to the lower layer, $\delta x > 0$, the particles in the upper layer fall into the gaps between the particles in the layer below and the volume occupied by the "soil" reduces $\delta z > 0$. The relationship between horizontal movement (shear displacement) and vertical movement (volume change) is shown in Fig. 8.9c (for this ideal packing of equally sized circular discs, it is part of a cosine curve). On the other hand, Fig. 8.9b shows an initially dense packing of circular particles in two layers. Now, for the upper layer to move sideways relative to the lower layer, the particles in this layer are forced to climb over the particles in the underlying layer and the volume occupied by the "soil" increases, $\delta z < 0$ (Fig. 8.9c). The particles in a real soil are arranged in a much more complex way than that shown in Fig. 8.9 but we can certainly anticipate that – and not be surprised that – volume changes may accompany shearing of a real granular material. This phenomenon is called *dilatancy* and the extent of the volume change that occurs during

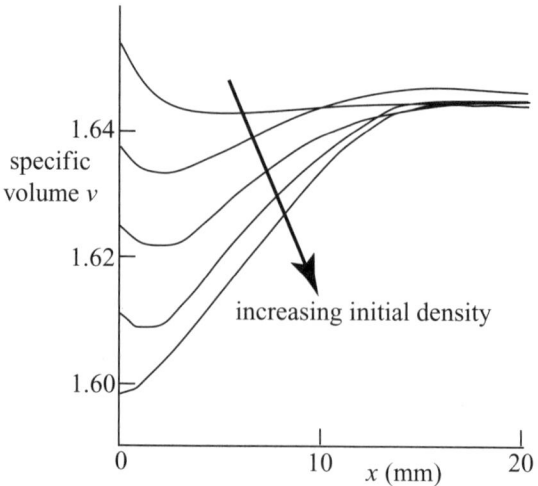

Figure 8.10. Shearing of samples of steel balls with different initial densities; normal stress $\sigma'_z = 138$ kPa (adapted from Wroth[4]).

shearing is expected to be strongly influenced by the density of the packing. Osborne Reynolds (of Reynolds' number fame) noted that dilatancy is well-known to those who buy and sell corn: you are recommended to buy such bulk materials by mass rather than by volume.

Results from shearing tests on steel balls (a slightly closer analogue for a real granular material) are shown in Fig. 8.10.[4] These show the changes in specific volume v with shear displacement x for samples prepared with different initial densities (or specific volumes) but with the same normal stress. We see first that, just as we have proposed, the dilatancy – the volume change accompanying shearing – is indeed very dependent on the initial density: the densest sample shows almost nothing but expansion whereas the loosest sample shows almost nothing but compression. Second, we see that, irrespective of the initial specific volume or density, all the samples end at the same value of specific volume as the shear displacement is increased. This is exactly what is implied by our strength model (Fig. 8.8). In simple terms, if the soil does not currently have the correct density or specific volume for failure with the current normal effective stress, then it has to change its specific volume in order that it may eventually reach the correct value. The results of many shear tests on steel balls, including those in Fig. 8.10, are presented in the form of our strength model in Fig. 8.11 in three plots of (a) shear stress and normal stress, (b) normal stress and specific volume, and (c) shear stress and specific volume.

In our discussion of the effect of changing the stress on a layer of soil in Chapter 6, we observed that change in volume required movement of pore fluid into or out of the voids of the soil and that, if the permeability of the soil to the flow of the pore fluid (typically water) were low, it would be difficult for any volume change to

---

[4] Wroth, C.P. (1958) Soil behaviour during shear – existence of critical voids ratios. *Engineering* **186** 409–413.

8.6 Drained and undrained strength

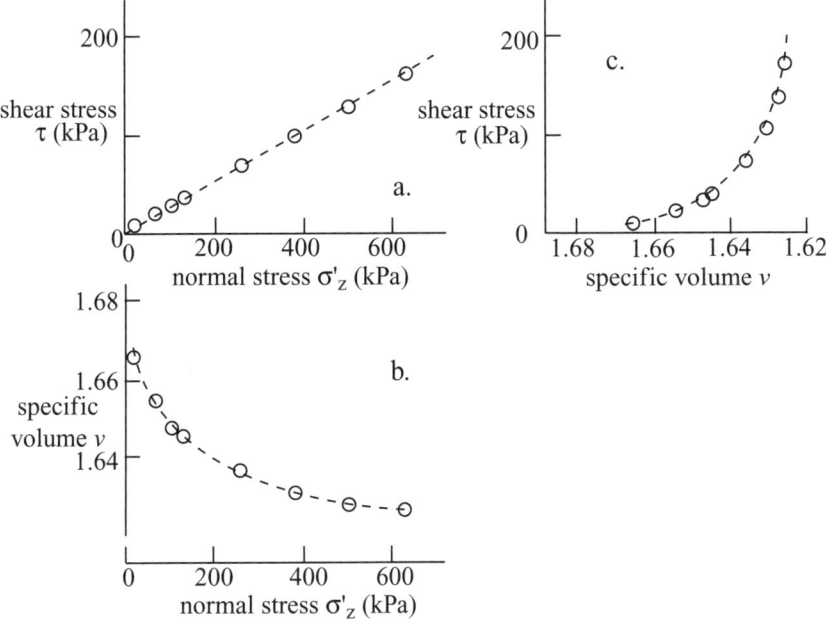

Figure 8.11. Failure conditions from shearing tests on steel balls (adapted from Wroth[4]).

occur rapidly and it might be necessary for pore water pressures to develop, at least temporarily, to change the effective stress and allow the soil to react to the changed external conditions without changing in volume. Exactly the same sort of argument can be made in the context of dilatancy and strength.

For a high permeability soil (sand or gravel), the water can usually flow freely, and volume change will accompany the shearing with no change in pore pressure from the ambient equilibrium value. As previously, we can describe this as the *drained* response (Section 6.2).

For a low permeability soil (clay), the pore water cannot flow freely. The strength that can be generated according to our strength model (Fig. 8.8) is governed by the initial volumetric packing, which implies a particular associated effective normal stress σ'_z at failure. If the current effective stress is different from this particular value, the pore pressure has to change to produce that value: this is the *undrained* response. Of course, in the long term we can expect that the pore pressure will dissipate and reach some equilibrium condition. With the accompanying flow, the volume will change and the eventual strength τ_f will be the strength associated with the original effective normal stress – or whatever the equilibrium effective normal stress has now become.

8.6 Drained and undrained strength

Let us put some symbols and then some numbers into this discussion. Suppose that we have a soil which has a frictional strength given by:

$$\tau_f = \sigma'_{zf} \tan \phi \tag{8.2}$$

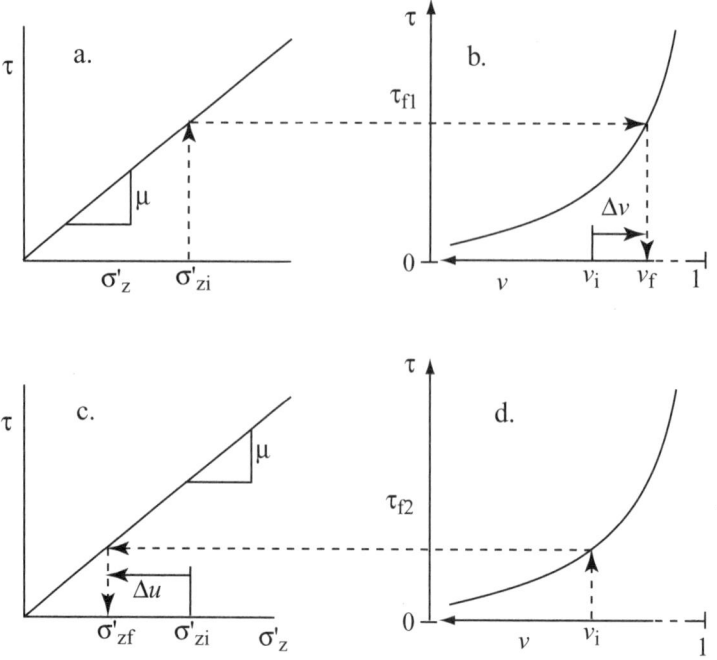

Figure 8.12. (a, b) Drained strength; (c, d) undrained strength.

Let us suppose that the link between strength and specific volume at failure is described by an exponential relationship:

$$v_f = v_{min} + \Delta v \exp\left\{-(\tau_f/[\sigma_{ref} \tan \phi])^\beta\right\} \qquad (8.3)$$

where σ_{ref} is a reference stress, v_{min} and Δv define a range of available specific volumes over which the soil can exist, and β is a soil constant controlling the link between change in volumetric packing and development of strength. Such an expression reflects our broad expectation that the more densely packed the soil becomes, the more rapidly will the strength increase. In principle, the strength becomes infinite when $v = v_{min}$, which might well correspond to zero voids, $v_{min} \approx 1$, by which stage the soil is no longer a soil. At the other extreme, when $v = v_{min} + \Delta v$, the strength is zero: this is the loosest packing in which the soil can barely transmit stress. The present effective normal stress and specific volume are σ'_{zi} and v_i, respectively.

Keeping the effective normal stress constant, at failure $\sigma'_{zf1} = \sigma'_{zi}$ and the available strength is $\tau_{f1} = \sigma'_{zi} \tan \phi$ and in order that this strength may be mobilised the specific volume will have to change to:

$$v_{f1} = v_{min} + \Delta v \exp\left\{-(\tau_{f1}/[\sigma_{ref} \tan \phi])^\beta\right\} \qquad (8.4)$$

and this describes the *drained* strength (Figs 8.12a, b). The strength is controlled by the vertical effective stress from (8.2) and this strength is associated with a particular density of packing from (8.3). If the initial density does not match with this eventual

8.7 Clay: overconsolidation and undrained strength

density it has to change, and the density change implies movement of pore fluid (water, if the soil is saturated) and hence drainage of the soil.

On the other hand, if we keep the specific volume constant so that at failure $v_{f2} = v_i$, the strength that we can mobilise is, by inversion of (8.3):

$$\tau_{f2} = \sigma_{ref} \tan \phi \left[\ln \left(\frac{\Delta v}{v_i - v_{min}} \right) \right]^{1/\beta} \tag{8.5}$$

and the effective normal stress must change to $\sigma'_{zf2} = \tau_{f2}/\tan\phi$ in order that it may become compatible with this strength. If the total normal stress remains constant, the pore pressure must change by an amount $\Delta u = \sigma'_{zi} - \sigma'_{zf2}$ (Fig. 8.12c, d). We are not permitting any movement of pore fluid, and the process of development of this strength is consequently described as *undrained*.

Suppose we have a soil with an angle of shearing resistance $\phi = 25°$ and with $\sigma_{ref} = 100$ kPa, $v_{min} = 1.1$, $\Delta v = 0.9$, $\beta = 0.6$; and suppose we have initial conditions: specific volume $v_i = 1.6$, vertical total stress $\sigma_{zi} = 75$ kPa, pore pressure $u = 0$, to give vertical effective stress $\sigma'_{zi} = 75$ kPa. In the *drained* case, the strength is governed by the effective normal stress so that the eventual strength is $\tau_{f1} = \sigma'_{zi} \tan \phi = 75 \tan 25° = 35$ kPa. The corresponding value of specific volume is $v_f = v_{min} + \Delta v \exp\{-(\tau_f/[\sigma_{ref}\tan\phi])^\beta\} = 1.1 + 0.9 \exp\{-(35/[100\tan 25])^{0.6}\} = 1.488$ and the change of volume accompanying the shearing to this ultimate strength is $\Delta v = 1.488 - 1.6 = -0.112$, which implies a compression of the soil, an increase in density.

In the *undrained* case, the strength is governed by the initial volumetric packing so that this strength is, from (8.5):

$$\tau_{f2} = 100 \times \tan 25° [\ln(0.9/(1.6 - 1.1))]^{1/0.6} = 19.2 \text{ kPa} \tag{8.6}$$

Mobilisation of such a strength implies an effective normal stress at failure $\sigma'_{zf2} = \tau_{f2}/\tan\phi = 19.2/\tan 25° = 41.2$ kPa. The required change in effective normal stress is thus $\Delta u = \sigma'_{zi} - \sigma'_{zf2} = 75 - 41.2 = 33.8$ kPa.

The calculation logic is thus: for high permeability soil: $\sigma'_{zi} \to \tau_{f1} \to v_f$ (the strength is controlled by the vertical effective stress and the density adjusts accordingly); and for low permeability soil: $v_i \to \tau_{f2} \to \sigma'_{zf} \to \Delta u$ (the strength is controlled by the density and the vertical effective stress adjusts accordingly). For low permeability soil the undrained strength is often given the symbol c_u or s_u.

8.7 Clay: overconsolidation and undrained strength

Overconsolidation was seen in Section 4.7 to be a route by which reductions in specific volume (increases in density) could be locked into the soil. Figure 8.13 illustrates this phenomenon. The dotted curve AC shows the volumetric compression on original loading in an oedometer to the maximum, preconsolidation pressure σ'_{zmax} at point C on the normal compression line. The vertical stress is then reduced to the current, initial stress σ'_{zi} at point I with specific volume v_i. The unloading stiffness of soils is usually considerably higher than the original stiffness in "virgin" loading

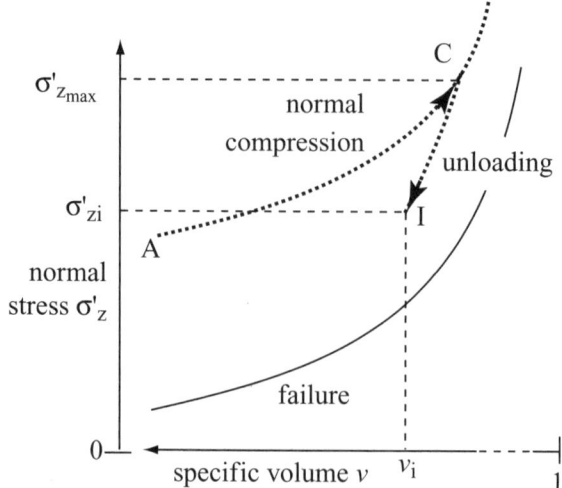

Figure 8.13. Overconsolidation: vertical effective stress and specific volume.

(compare AC and CI, Section 4.7). The overconsolidation ratio, the ratio of past maximum to present vertical effective stress, $n = \sigma'_{z_{max}}/\sigma'_{zi}$, characterises the extent to which the load, historically applied, has been removed.

The solid curve in Fig. 8.13 represents the link between specific volume and vertical effective stress at failure. The curves for normal compression and for failure in Fig. 8.13 have been drawn with a similar shape but this is not necessarily the case: see Exercise 7 in Section 8.13. The non-coincidence of these two curves indicates the need for adjustment of effective stress or specific volume during shearing to failure. The curves from Fig. 8.13 have been reproduced in Fig. 8.14c together with the link between shear stress τ and specific volume v at failure (Fig. 8.14b) and the link between shear stress τ and normal effective stress σ'_z at failure (Fig. 8.14a).

The density of the soil – and hence the current strength of the soil – is primarily determined by the past maximum effective stress (Figs 8.13, 8.14c). The current density or specific volume v_i determines the undrained strength $\tau_f = c_u$ (Fig. 8.14b), which then determines the effective normal stress at failure σ'_{zf} (Fig. 8.14a). The density change on unloading and reloading of overconsolidated clays is probably relatively small. It is not unreasonable to suggest a link between the compression and strength behaviour for clays such that the undrained strength $\tau_f = c_u$ is related to the maximum precompression stress, or preconsolidation pressure $\sigma'_{z_{max}}$, by a simple empirical relationship:

$$c_u = \Lambda \sigma'_{z_{max}} = \Lambda n \sigma'_{zi} \approx 0.2 n \sigma'_{zi} \qquad (8.7)$$

With knowledge of *in-situ* stress and overconsolidation ratio, we can produce an estimate of the undrained strength. Such an estimate can be useful for initial studies of design feasibility before more direct data become available from which the actual strength of the soil can be directly discovered.

8.8 Pile load capacity

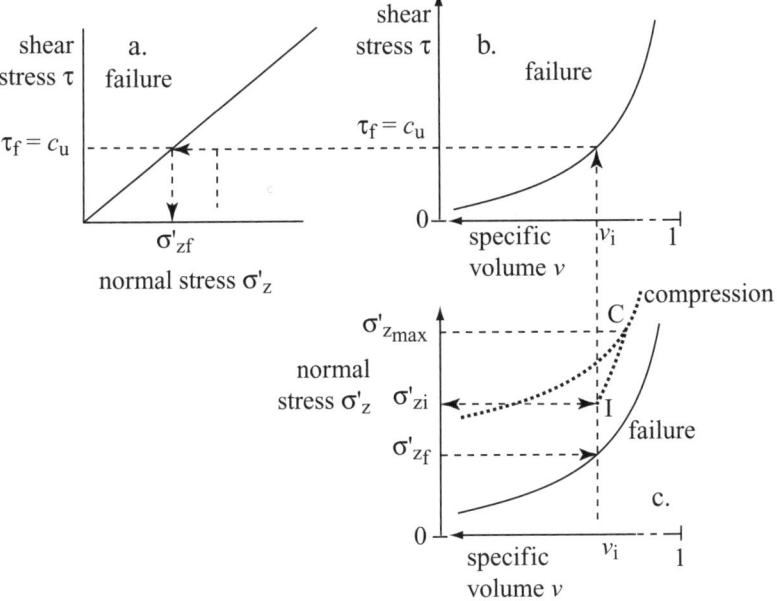

Figure 8.14. Overconsolidation and undrained strength.

8.8 Pile load capacity

The estimation of the load capacity of a pile provides a simple example of the application of the strength model. Let us suppose that we have a long slender pile of length ℓ being loaded axially (Fig. 8.15). We will assume that the pile is long enough and slender enough that all the load is transferred to the surrounding soil through shaft resistance – that is, through the mobilisation of shear stresses down the length of the pile, so that the base resistance σ_b in Fig. 8.1 is negligible. The vertical

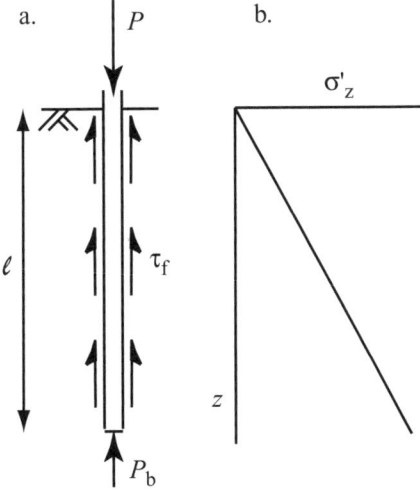

Figure 8.15. Pile foundation, load transfer through shaft resistance.

effective stress in the ground varies linearly with depth $\sigma'_z = \rho g z$. Let us suppose, first, that we are concerned with a pile in dry sand so that we are not worried about pore pressures. We are trying cautiously to retain our restriction to a single dimension in our applications so, while it is reasonable to propose that this vertical effective stress gives a general indication of the level of stress in the soil around the pile, we cannot really insist that it is exactly the normal stress pushing on the shaft of the pile.

We saw in Section 4.4.1 that, according to elastic theory, in one-dimensional compression the soil will push laterally onto its container with a stress which is a proportion $\nu/(1-\nu)$ of the vertical stress, where ν is Poisson's ratio. Our estimate of the available shear stress on the side of the pile when it is failing by penetrating slowly but inexorably into the ground is then:

$$\tau_f = \frac{\nu}{1-\nu}\sigma'_z \tan\phi = \frac{\nu}{1-\nu}\rho g z \tan\phi \tag{8.8}$$

If the pile has perimeter s, the ultimate load for this axial failure of the pile is obtained by integrating this shear resistance over the length of the pile:

$$P_u = \frac{\nu}{(1-\nu)}\rho g s \tan\phi \int_0^\ell z \, dz = \frac{\nu}{2(1-\nu)}\rho g s \ell^2 \tan\phi \tag{8.9}$$

In a dry sand with density 1.7 Mg/m^3, angle of shearing resistance $\phi = 30°$ ($\tan\phi = 0.58$) and Poisson's ratio $\nu = 0.2$, a circular pile of length $\ell = 15$ m and diameter $2r = 0.3$ m (perimeter $s = 2\pi r = 0.94$ m) will have an axial capacity $P_u = 0.25 \times 1.7 \times 9.81 \times 0.94 \times 15^2 \times 0.58/2 = 255$ kN.

Piles are deliberately much stiffer than the ground through which they are driven or inserted. They are used precisely to transmit the loads from the surface, where the stress level in the soil is definitely low and the strength probably low. But other things may be happening on the site around the pile foundations – for example, some site preparation such as raising the ground level by placement of a layer of fill (as in the examples in Section 6.3). Such filling increases the total stresses in the ground and causes the ground to compress – such compression occurring slowly if the ground is a low permeability clay. If the foundation pile is sitting on an underlying hard layer, the settling ground will generate downward shear stresses on the shaft of the pile – *down-drag* – which act in the same direction as the load applied at the surface P_t (Fig. 8.16). So far as the stresses in the pile are concerned, which will govern its structural design – the necessary strength of the concrete, for example – the largest stress is not at the ground surface, where the load is applied, as might initially be supposed.

Shaft resistance τ_f develops with relative movement between pile and soil. As a second example, consider a pile in normally compressed clay in which the maximum shaft resistance that can be developed can be related to the *in-situ* vertical effective stress σ'_z through a constant Λ (8.7).

The ultimate load of an end-bearing pile of length ℓ and radius r (perimeter $s = 2\pi r$) in consolidating clay (Fig. 8.16) is calculated assuming a failure mechanism in which the pile moves steadily downwards and is made up from a combination of

8.8 Pile load capacity

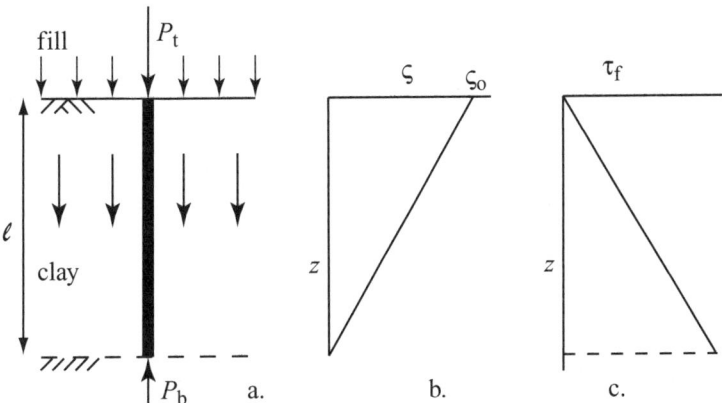

Figure 8.16. Pile foundation: (a) down-drag from settling ground; (b) variation of soil settlement with depth; (c) assumed full mobilisation of available shear strength at all depths.

the shaft shear stresses, producing a force P_s – which would be generated with the pile moving down relative to the soil – and the end bearing at the toe of the pile, generating a force P_b, so that the ultimate load is $P_f = P_s + P_b$, and in this failure condition the force remaining in the pile at its toe is $P_f - P_s$. The shaft resistance is:

$$P_s = \pi \Lambda \rho' g r \ell^2 \tag{8.10}$$

assuming that vertical effective stress is being generated uniformly with depth in a saturated soil with density ρ and buoyant density $\rho' = \rho - \rho_w$.

Under working or serviceability conditions, on the other hand, the direction of relative movement between the soil and the pile is reversed, and the soil moving downwards generates an additional load ΔP at the toe of pile given by:

$$\Delta P = \int_0^\ell 2\pi r \tau \, dz = 2\pi r \Lambda \int_0^\ell \sigma'_z \, dz \tag{8.11}$$

giving

$$\frac{\Delta P}{P_s} = 1 \tag{8.12}$$

and the stress at the toe of the pile is considerably increased.

Let us assume that we have a pile of length $\ell = 15$ m and diameter $2r = 0.3$ m in clay with density $\rho = 1.5$ Mg/m^3 and $\Lambda = 0.2$. The pile capacity $P_s = 104$ kN and the extra load arising from the down-drag will similarly be 104 kN over the cross section of the pile of diameter 0.3 m, representing an extra stress of 1.47 MPa.

The ultimate load capacity of the pile is not affected by down-drag. If the pile is failing, the mechanism of failure implies sufficient downward movement of the pile to generate full positive shaft shear stresses resisting the movement. However, if the pile design has overlooked the contribution to axial load that the settling surrounding soft soil may provide, there may be insufficient relative downward movement of the pile to ensure that the shear stresses round the perimeter of the pile are helping to reduce the axial force in the pile rather than increasing it.

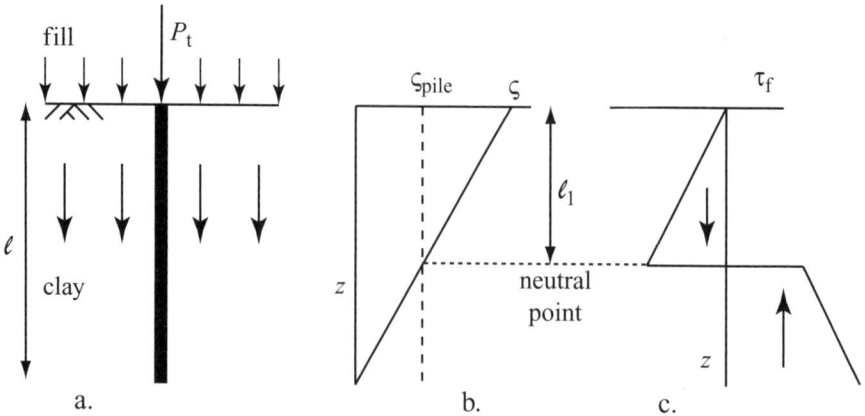

Figure 8.17. (a) Down-drag of "floating" pile; (b) settlement of soil ς matches settlement of pile ς_{pile} at neutral point at depth ℓ_1; (c) variation with depth of shear stress at pile:soil interface.

A slightly more complicated configuration is obtained when considering "floating" friction piles (Fig. 8.17). Now there is no end bearing contribution to the capacity of the pile, $P_b = 0$, and under any working loading the contributions from the shear stresses around the pile must combine with the applied load P_t at the ground surface to give overall vertical equilibrium. The pile will move downwards by some distance ς_{pile} (Fig. 8.17b) so that there is a neutral point at depth ℓ_1 at which there is no relative displacement between the pile and the soil, and at which the direction of the shaft shear stress reverses (Fig. 8.17b, c). We assume that shaft shear stresses – either upwards or downwards – always fully mobilise the available strength. We also assume that the pile is rigid, or at least much stiffer than the settling soil so that deformation of the pile can be neglected.

Vertical equilibrium requires that the downward forces on the upper part of the pile, down to depth ℓ_1, should exactly balance the upward forces on the lower part of the pile, between depths ℓ_1 and ℓ:

$$P_t + \int_0^{\ell_1} \Lambda 2\pi r \rho' g z \, dz = \int_{\ell_1}^{\ell} \Lambda 2\pi r \rho' g z \, dz \qquad (8.13)$$

giving:

$$P_t + \pi \Lambda r \rho' g \ell_1^2 = \pi \Lambda r \rho' g \left(\ell^2 - \ell_1^2 \right) \qquad (8.14)$$

so that:

$$\left(\frac{\ell_1}{\ell} \right)^2 = \frac{1}{2} \left(1 - \frac{P_t}{\pi \Lambda \rho' g r \ell^2} \right) = \frac{1}{2} \left(1 - \frac{P_t}{P_f} \right) \qquad (8.15)$$

where $P_f = \pi \Lambda \rho' g r \ell^2$ is the overall capacity of this friction pile (8.10).

The variation of axial load in the pile is then, for $0 < z < \ell_1$:

$$\frac{P}{P_f} = \frac{P_t}{P_f} + \left(\frac{z}{\ell} \right)^2 \qquad (8.16)$$

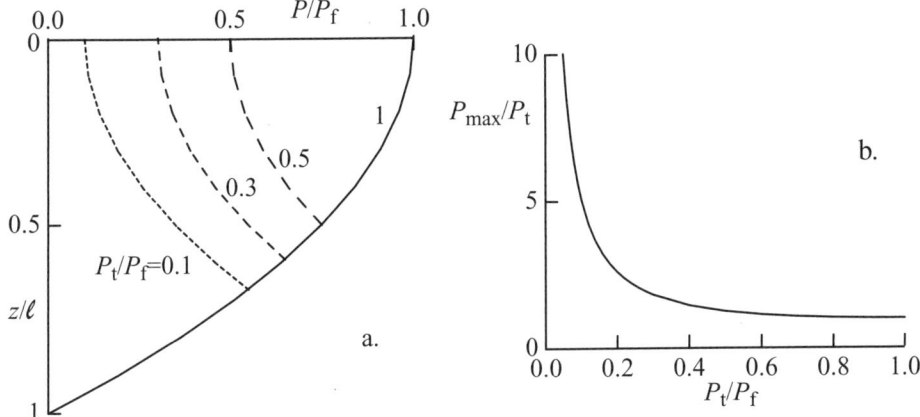

Figure 8.18. (a) Axial load distribution in floating piles; (b) magnification of axial load as function of load applied at top of pile.

and for $\ell_1 < z < \ell$:

$$\frac{P}{P_f} = \frac{P_t}{P_f} + 2\left(\frac{\ell_1}{\ell}\right)^2 - \left(\frac{z}{\ell}\right)^2 = 1 - \left(\frac{z}{\ell}\right)^2 \qquad (8.17)$$

The variation of axial load is shown in Fig. 8.18a for different values of P_t/P_f.

The ultimate capacity of the pile is not affected. If the pile is moving downwards sufficiently far to generate positive shaft resistance, the limiting curve in Fig. 8.18a corresponds to $P_t/P_f = 1$ and the axial load falls parabolically from the top of the pile. However, if the design of the pile foundation deliberately chooses the dimensions of the pile such that the applied load at the top of the pile P_t uses only a fraction of the available shaft resistance, the maximum load in the pile, P_{max}, which will occur at the neutral point $z = \ell_1$, may be greatly magnified. The variation of this magnification ratio P_{max}/P_t with apparent pile load factor P_t/P_f can be found by combining (8.15) and (8.16):

$$\frac{P_{max}}{P_t} = \frac{1}{2}\left(\frac{P_f}{P_t} + 1\right) \qquad (8.18)$$

This result is plotted in Fig. 8.18b. If a reduction factor of 3 is placed on the load at ground level, $P_t/P_f = 1/3$, the maximum force in the pile is $P_{max} = 2P_t$. The input load at the ground surface P_t may provide only a very unsafe estimate of the maximum axial load in the pile.

8.9 Infinite slope

The analysis of failure of slopes using mechanisms such as those shown in Fig. 8.3 really takes us away from our one-dimensional environment. However, there may be good reasons why, in some circumstances, failure of a slope occurs on a plane

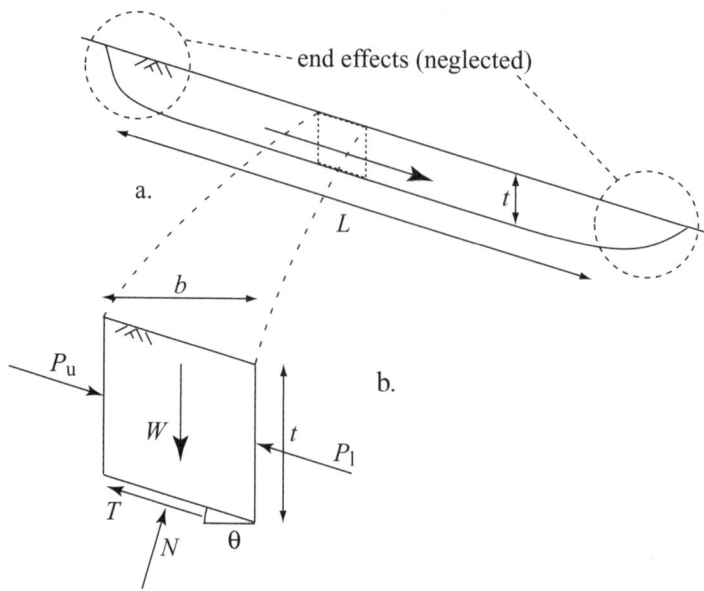

Figure 8.19. Shallow slope failure: (a) "infinite" slope; (b) element of soil from failure mechanism.

parallel to the slope and over a long distance down the slope (Fig. 8.19a). There will be end effects at the top and bottom of the failing slope – within the dotted circles in Fig. 8.19a – but, if the depth t to the failure surface is small in comparison with the length of the failing soil, $t \ll L$, we can ignore the ends and we can invoke symmetry to permit a simple analysis.

An element of width b (and thickness d into the diagram) from the failing slope is shown in Fig. 8.19b. The forces acting on this element are its weight W, normal N and tangential (shear) T forces on the base of the element, and forces P_u and P_l on the vertical sides of the element, up-slope and down-slope. The slope is long and the failure is occurring parallel to the slope, so we can take our element from any position we choose within the failing soil and the conditions will be the same no matter where this element is located. It follows therefore that the forces acting on the up-slope and down-slope sides of the element must exactly balance so we can ignore them in our analysis of the equilibrium of the element: the weight is entirely equilibrated by the normal and shear forces on the base. The weight of the element is:

$$W = \rho g b t d \qquad (8.19)$$

Equilibrium orthogonal to the slope then gives:

$$N = W \cos \theta \qquad (8.20)$$

and:

$$T = W \sin \theta \qquad (8.21)$$

8.9 Infinite slope

We can convert the normal force N and shear force T to corresponding normal and shear stresses, noting that the cross-sectional area of the base of the element is $bd/\cos\theta$:

$$\sigma_n = \frac{N\cos\theta}{bd} \qquad (8.22)$$

$$\tau = \frac{T\cos\theta}{bd} \qquad (8.23)$$

Rearranging we find:

$$\sigma_n = \rho g t \cos^2\theta \qquad (8.24)$$

and:

$$\tau = \rho g t \cos\theta \sin\theta \qquad (8.25)$$

In general, there will also be a pore pressure u and corresponding equivalent force $U = ubd/\cos\theta$ on the base of the element, so that the effective normal stress is correspondingly reduced to $\sigma'_n = \sigma_n - u$. There is some benefit to be obtained by relating the pore pressure to the total vertical stress through a pore pressure ratio $r_u = u/\rho g t$, so that the effective normal stress can be written $\sigma'_n = \rho g t(\cos^2\theta - r_u)$. The mobilised friction on the base of the soil element is:

$$\tan\phi_{mob} = \frac{\tau}{\sigma'_n} = \frac{\cos\theta\sin\theta}{\cos^2\theta - r_u} = \frac{\tan\theta}{1 - r_u\sec^2\theta} \qquad (8.26)$$

There are various cases that we can consider. If we suppose that the slope is dry, so that there are no pore water pressures, $r_u = 0$, the normal stress in (8.24) is an effective stress $\sigma_n = \sigma'_n$, and the mobilised angle of friction, ϕ_{mob}, for the base of the soil element is given by:

$$\tan\phi_{mob} = \tau/\sigma'_n = \tan\theta \qquad (8.27)$$

or $\phi_{mob} = \theta$, and we can say that the slope will be stable provided $\theta < \phi$, where ϕ is the available angle of shearing resistance.

A second possibility would be a slope in clay in which a rapid failure is controlled by the undrained strength c_u. Now the normal stress is not important – pore pressures will change as we have seen in Section 8.6 to guarantee that the correct combination of effective normal stress, shear strength and density is reached for failure of the soil. If we know that the failure is occurring at a depth t, our equation of stability requires:

$$\tau = \rho g t \cos\theta \sin\theta < c_u \qquad (8.28)$$

which becomes an equation for a limiting slope angle given knowledge that failure is likely to occur at a certain depth t:

$$\theta = \frac{1}{2}\sin^{-1}\left[\frac{2c_u}{\rho g t}\right] \qquad (8.29)$$

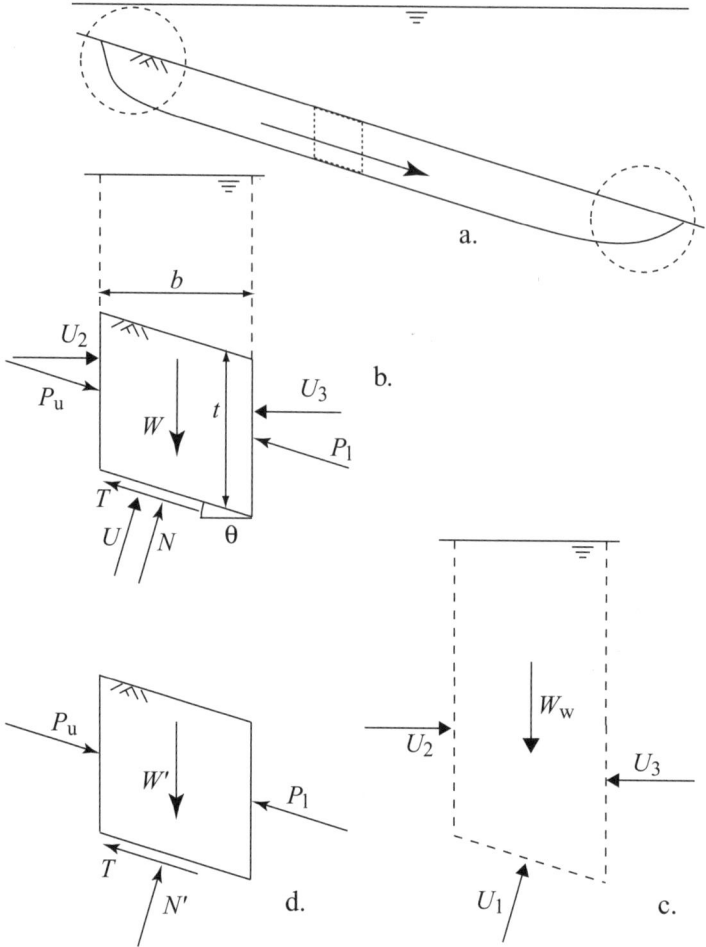

Figure 8.20. Submerged infinite slope.

Or, turned the other way round, if we know that a clay slope with a particular inclination is failing along a surface at a particular depth, we can estimate the *in-situ* undrained strength.

A third possibility would be a slope submerged in water but, as in our discussion of seepage (Section 5.2), not having any variations in total head and consequently not having any seepage flow (Fig. 8.20). We can conveniently separate the water and the submerged soil as shown in Figs 8.20c and 8.20d, respectively. The water pressures around the sides of the element (which now extends up to the water surface, Figs 8.20b, c) will balance the weight of the water column so that the water forces U_1 on the base and U_2 and U_3 on the sides of the element in Fig. 8.20c provide the necessary support for the column of water of weight W_w (calculated in just the same way as in Section 2.7). Then the submerged element of soil in Fig. 8.20d can be treated directly in terms of effective stresses, and the weight is now the buoyant or submerged weight W'. The analysis is unchanged from the analysis for the dry slope – the only difference is that we have to use the buoyant density $\rho' = \rho - \rho_w$ in

8.9 Infinite slope

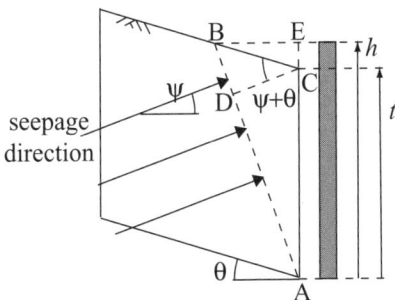

Figure 8.21. Infinite slope with steady seepage flow.

calculating the effective normal and shear stresses on the base of the soil element. The application of the criterion for stability of the slope produces the same equation as for the dry slope (8.27) and we again expect the slope to be stable provided $\theta < \phi$.

Other possibilities involving steady seepage in the slope can, with a little care, also be treated as one-dimensional cases. For seepage to occur, there must be a gradient of total head and there can be no flow between points which have the same total head (Section 5.2). If seepage is occurring in one direction (the one-dimensional assumption), then along any line orthogonal to the direction of seepage flow the total head must be constant: there is no flow in this direction, and therefore no gradient of total head.

Figure 8.21 shows a general configuration for our infinite slope. Seepage is occurring in a direction which makes an angle ψ with the horizontal, or $\psi + \theta$ with the slope. We are interested in finding the pore pressure on the base of our element, so we draw a line orthogonal to the seepage direction from the corner A. This intersects the slope at point B where the pore pressure must be zero because we have emerged into the open air. We need to find the height of B above A, or the length AE, indicated as h in Fig. 8.21, in terms of the depth t of the element. Working with the right-angled triangles ACD, BCD and BCE, in turn we can find:

$$h = t \left[1 + \frac{\sin \psi \sin \theta}{\cos (\psi + \theta)} \right] = t \frac{\cos \theta \cos \psi}{\cos (\psi + \theta)} \qquad (8.30)$$

The total head is constant along the line AB because it is a line orthogonal to the direction of flow, so the pressure head at any point along AB will be equal to the drop in elevation from point B (Section 5.2). The pressure head at A is thus h and the pore pressure at the base of our soil element is $u = \rho_w g h$, or:

$$r_u = \frac{u}{\rho g t} = \frac{\rho_w}{\rho} \frac{\cos \theta \cos \psi}{\cos (\psi + \theta)} \qquad (8.31)$$

Then the mobilised friction on the base of the element is:

$$\tan \phi_{mob} = \frac{\tau}{\sigma_n - u} = \frac{\rho \cos \theta \sin \theta}{\rho \cos^2 \theta - \rho_w \cos \theta \cos \psi / \cos (\psi + \theta)}$$
$$= \frac{\rho \tan \theta}{\rho' - \rho_w \tan (\psi + \theta) \tan \theta} \qquad (8.32)$$

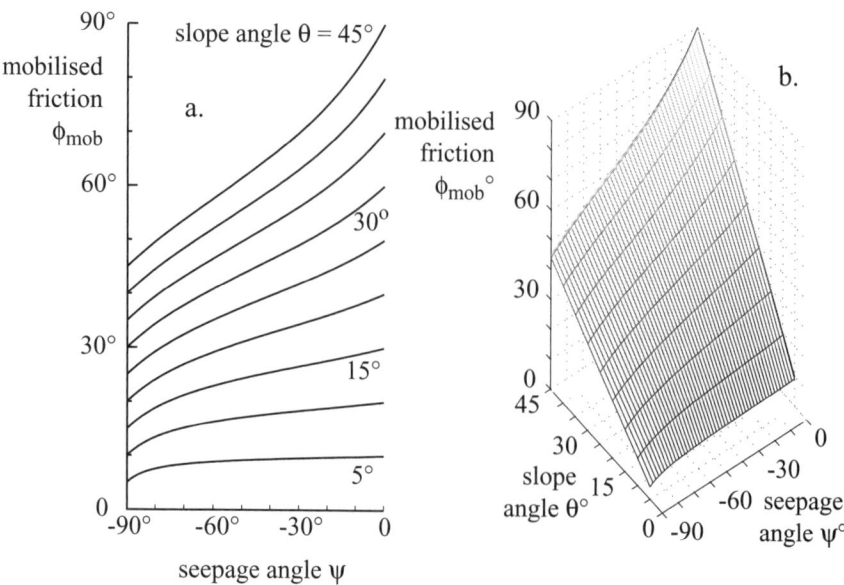

Figure 8.22. Mobilised friction ϕ_{mob} in infinite slope (angle θ) with steady seepage at ψ to horizontal (drawn for $\rho = 2\rho_w = 2\rho'$).

where $\rho' = \rho - \rho_w$ is the buoyant density of the soil. This relationship is shown in Fig. 8.22. Various simple situations are shown in Fig. 8.23.

Flow parallel to the slope (Fig. 8.23a) is a common situation since it is quite likely that the same geological features that influence the formation of a failure plane parallel to the slope will also influence the internal flow regime. This implies that $\psi = -\theta$ and the pore pressure on the base of the element is $u = \rho_w gt \cos^2 \theta$, $r_u = (\rho_w/\rho) \cos^2 \theta$. The mobilised angle of friction is:

$$\tan \phi_{mob} = \frac{\rho}{\rho'} \tan \theta \qquad (8.33)$$

We recall from Section 3.3.3 that the ratio of saturated bulk density to buoyant density is:

$$\frac{\rho}{\rho'} = \frac{G_s + e}{G_s - 1} \qquad (8.34)$$

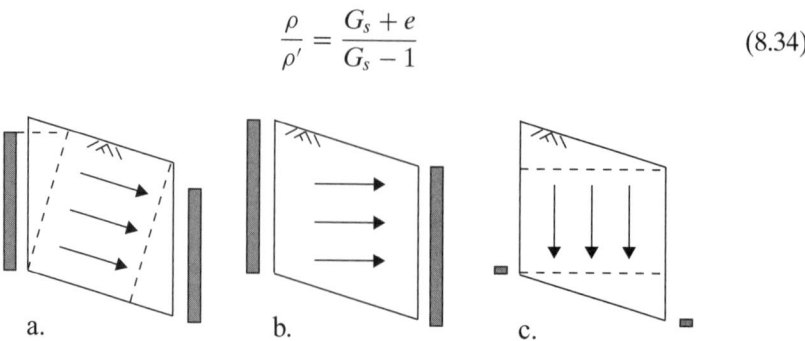

Figure 8.23. Element from infinite slope with (a) seepage parallel to the slope; (b) horizontal seepage; (c) seepage vertically downwards. In each case, the bars indicate the pressure head at the base of the soil element.

8.9 Infinite slope

where the specific gravity of the soil mineral G_s is probably in the range 2.6 to 2.7, and the void ratio e might be anything in the range of 0.3 to 1 (or more). Even with this range of void ratios, the variation of ρ/ρ' is only from about 1.8 to 2.1 so we can say as a generalisation that it will be of the order of 2 and the mobilised angle of friction is of the order of twice the slope angle. The stable angle of the slope will be of the order of $\phi/2$, where ϕ is the available frictional strength of the soil. The presence of seepage parallel to the slope thus has a very damaging effect on the stability of the slope.

Figure 8.23b shows seepage occurring horizontally, perhaps driven by some underlying impermeable horizontal layer. The pore pressure head on the base of the element is equal to the depth t: the line orthogonal to the direction of flow goes straight up to meet the free surface with $h = t$. For this case, $\psi = 0$ and the mobilised angle of friction is:

$$\tan \phi_{mob} = \frac{\tan \theta}{1 - (\rho_w/\rho)\sec^2 \theta} \tag{8.35}$$

For $\rho/\rho_w = 2$, this implies that $\phi_{mob} = 2\theta$.

Finally, Fig. 8.23c shows downward vertical flow, which might arise as a result of infiltration of rainfall from the surface of the slope. A horizontal line (orthogonal to the direction of flow) through the bottom corner of the element reaches daylight at the same level as that corner, so the pressure head at the base of the element is zero. The flow is driven by gravity, without modification: the downward hydraulic gradient is $i = 1$. The downward flow produces a seepage body force as a result of the resistance to flow provided by the tortuous paths around and between the soil particles (Section 5.7). This seepage force (a body force of $\rho_w g i = \rho_w g$) leads to an increase in the apparent unit weight of the soil by an amount that exactly balances the Archimedes uplift. Fitting this information into the reference geometry of the section, we find $\psi = -\pi/2$ and the pore pressure on the base of the element, $u = 0$. We can calculate the mobilised angle of friction using (8.32):

$$\tan \phi_{mob} = \frac{\rho \tan \theta}{\rho' + \rho_w} = \tan \theta \tag{8.36}$$

and the condition for stability of the slope is the same as for the dry slope with no seepage: $\theta < \phi$.

We note that, although we have confined ourselves to a one-dimensional view of the world, we can nevertheless make rational statements about the stability or margin of safety for some quite realistic geotechnical systems.

8.9.1 Laboratory exercise: Angle of repose

There is a simple laboratory experiment that can be performed without any elaborate equipment to apply the understanding of slope stability that we have just gained and thus to estimate the angle of shearing resistance ϕ in the strength model of a granular soil (sand or fine gravel). All we need is a largish beaker, some sand, and a protractor for measuring angles (Fig. 8.24).

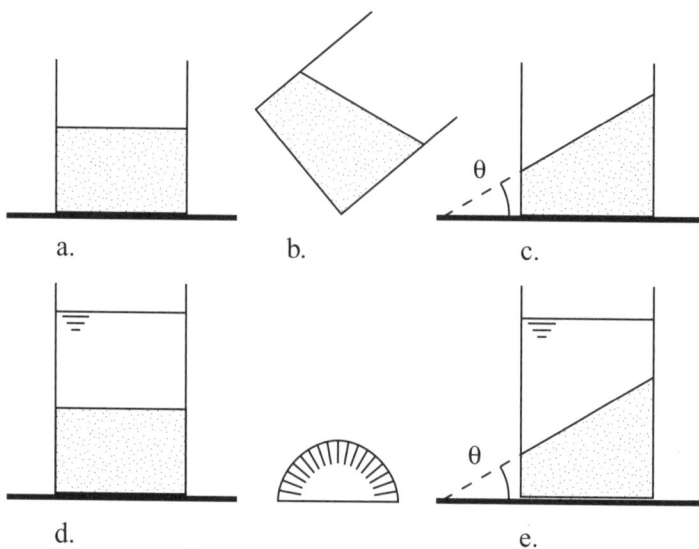

Figure 8.24. Laboratory experiment to estimate angle of repose: beaker of sand with protractor.

The sand is placed in the beaker (Fig. 8.24a) and the surface levelled off. The beaker is then tilted well beyond the point at which the sand surface becomes unstable (Fig. 8.24b) and then brought back to the horizontal (Fig. 8.24c). The surface of the sand is a small scale "infinite" slope, and the angle θ of the slope (Fig. 8.24c) gives an estimate of the angle of shearing resistance ϕ of the sand (8.27).

It is instructive to repeat the experiment with the sand in a beaker full of water (Fig. 8.24d). The beaker for this experiment may need to be somewhat taller than the one used for the test with dry sand to ensure that when the beaker is tilted and then brought back to the horizontal (Fig. 8.24e), the water does not overflow. The angle of repose of the submerged slope can be measured in the same way as for the dry slope and, according to our analysis, it should give the same result since, in the absence of seepage, the stable angle is ϕ whether the sand is in air or water (Figs 8.19, 8.20).

A further demonstration can be transferred to the beach (Fig. 8.25). We all know that to build sand castles or other sand sculptures, it is necessary to make use

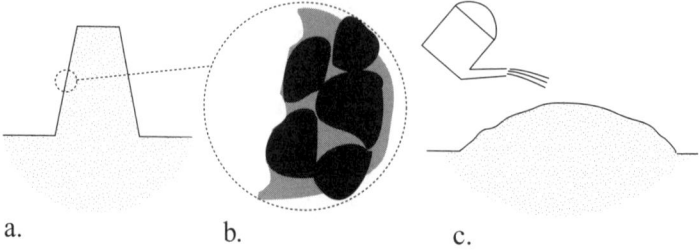

Figure 8.25. (a) Sand castle of damp sand; (b) menisci between sand particles; (c) water breaks down surface tension.

8.10 Undrained strength of clay: fall-cone test

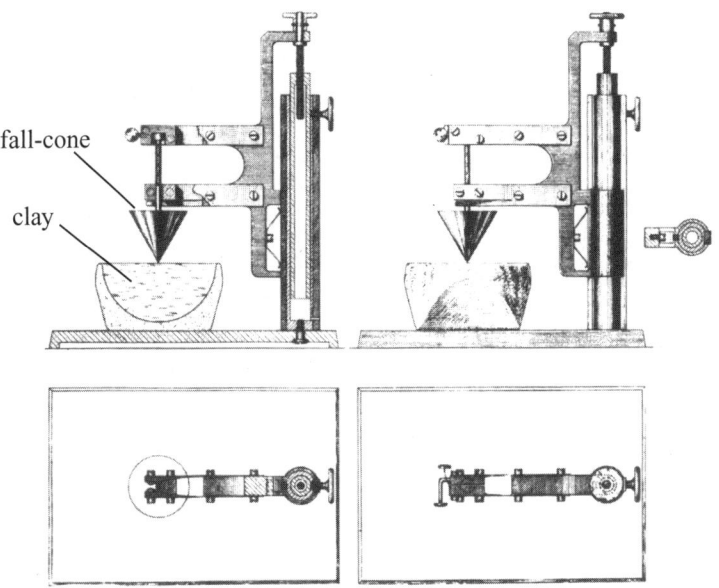

Figure 8.26. Fall-cone apparatus developed in Sweden.[5]

of damp sand. Dry sand cannot stand at an angle greater than the angle of repose, or the stable angle of an infinite slope, which for typical sands is around 30° to 40°. However, in damp sand there is a water meniscus between the sand particles (Fig. 8.25b) which, through the effect of surface tension (Section 2.8), produces a negative pore pressure inside the sand castle which, through the Principle of Effective stress, converts zero total stress (there is no stress acting on the steep surface of the sand castle) into a positive effective stress which is able to generate some frictional strength. If water is poured onto the sand castle (Fig. 8.25c), the surface tension breaks down and it is as if the sand were at best submerged and able only to stand at the angle of repose – or, since this is a transient process, at a shallower angle as a result of seepage occurring through the sand as it collapses.

8.10 Undrained strength of clay: fall-cone test

Undrained strength is a concept that we expect to meet when we are shearing less permeable soils (clays) when we load them too rapidly for the water to move through the pores and allow the clay to change in volume as it is sheared. There are various "multi-dimensional" tests which explore the undrained strength of clays, but we will give a brief description of a simple test that both reveals the link between strength and volumetric packing or density and also provides some sort of index test for classification of clays.

Penetration tests are used with metals to give a quick, almost non-destructive, indication of the strength of the material by measuring the size of the impression left by a standard indenter under a particular force. The *fall-cone test* does the same thing for clay soils: Fig. 8.26 shows the fall-cone apparatus devised to enable a rapid

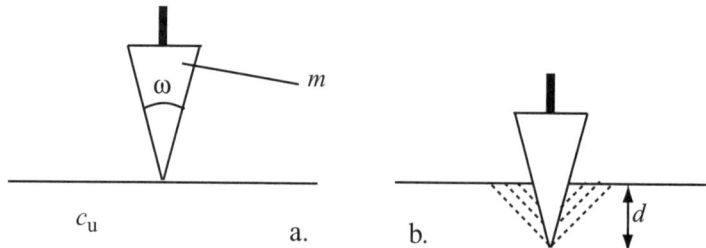

Figure 8.27. Fall-cone of mass m and tip angle ω allowed to fall under its own weight into clay of undrained strength c_u, coming to rest with penetration d.

estimate of the undrained shear strength of Swedish clays following a series of slope failure in railway cuttings.[5] A cone of mass m and with tip angle ω is allowed to fall under its own weight into a sample of clay from an initial point of contact with the level surface of the clay (Fig. 8.27). The penetration d of the cone is measured when it has come to rest. Now the way in which the clay strength is being mobilised around the tip of the cone as it penetrates is not as simple as the failure mechanisms that were shown in Figs 8.1, 8.3 and 8.4, but we can perhaps imagine sliding occurring on a series of surfaces as it moves (Fig. 8.27b). Whatever the detail of the mechanism, we can reasonably suggest that the penetration process will be controlled by the undrained strength c_u of the clay: the penetration occurs rapidly and drainage will not occur.

This then provides another example of the application of dimensional analysis to a geotechnical problem (see Section 3.6.2).[6] The variables in the cone test are the mass of the cone m, the tip angle ω, the undrained strength of the clay c_u, the eventual penetration d of the cone and, since the cone is falling under its own weight, the gravitational acceleration g. We are seeking dimensionless groups of these controlling variables and, in this case, the choice is limited. The building block dimensions are mass [M], length [L] and time [T]. The mass of the cone obviously has dimensions of mass [M] and the penetration has dimensions of length [L]. The tip angle is already dimensionless. The strength is a stress, a force per unit area, and force is mass × acceleration. Thus, the dimensions of undrained strength are [MLT^{-2}L^{-2}] or [ML^{-1}T^{-2}]. Gravitational acceleration has dimensions LT^{-2}. There are in fact only two dimensionless "groups" in the description and analysis of this problem. One is the dimensionless tip angle ω itself and the other is a ratio of the available undrained strength c_u to the stress generated over the cross-sectional area of the indentation (which is proportional to d^2) by the weight of the cone mg. This dimensionless ratio must be a function of ω:

$$\frac{c_u d^2}{mg} = f(\omega) = k_\omega \qquad (8.37)$$

---

[5] Statens Järnvägars Geotekniska Kommission 1914–1922 (1922) *Slutbetänkande avgivet til Kungliga Järnvägsstyrelsen*, Stockholm: Statens Järnvägar, Geotekniska Meddelanden 2.
[6] Palmer, A.C. (2008) *Dimensional analysis and intelligent experimentation*, World Scientific Publishing Company, Singapore.

8.11 Simple model of shearing ♣

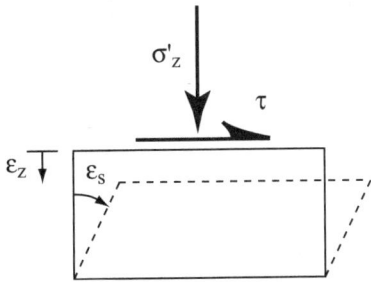

Figure 8.28. Element of soil for development of one dimensional model.

so that for a given cone angle ω, the dimensionless group $c_u d^2/mg$ should be a constant k_ω.

The fall-cone is used as a standard laboratory test to characterise the relationship between strength and density of packing for soils which can be treated as undrained when a cone is dropped into them. Studies have shown that, for a cone with $\omega = 30°$, the cone factor $k_{30} = 0.85$ and, for $\omega = 60°$, the cone factor $k_{60} = 0.29$ – these are standard cone tip angles.[7] With a cone mass of 80 g and cone angle 30°, a penetration $d = 20$ mm indicates that the soil under test has an undrained strength of $c_u = 1.7$ kPa. This is the standard configuration used to determine the so-called *liquid limit* of cohesive soils, which is actually the water content for which the soil has this particular strength, and thus provides a useful index for comparison of clays of different mineralogy.[8] A wider range of strengths can be measured by using cones of different masses or different angles: a heavier or sharper cone is needed to make a reliably detectable indentation in a strong clay; the more open 60° cone is useful for very weak clays.

8.11 Simple model of shearing ♣

We have accumulated sufficient ingredients through this book to be able to construct a complete model for the shearing of a sand (for example) in a shear box. We will idealise a little, but those who continue further with the study of the mechanics of soils will discover how this simple model can be developed into much more sophisticated models capable of being used in numerical analysis of complete and realistic prototype geotechnical problems.

We imagine an element of soil as shown in Fig. 8.28 which might be extracted from the central shearing region of the shear box in Fig. 8.6. It is subjected to a vertical, normal effective stress σ'_z and a shear stress τ. We expect that there will be vertical strains ε_z and shear strains ε_s. The shear strain produces a change in shape from rectangle to parallelogram, as indicated. Our task is to find a general link between changes in the stresses and changes in the strains.

---

[7] Wood, D.M. (1985) Some fall-cone tests. *Géotechnique* **35** 1, 64–68.
[8] In some countries, the standard configuration uses a cone of mass 60 g and tip angle of 60°, and seeks a penetration of $d = 10$ mm. This configuration also implies a strength $c_u = 1.7$ kPa.

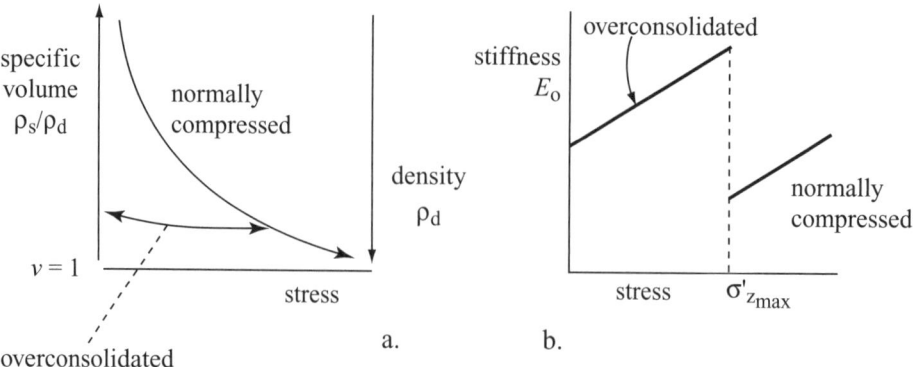

Figure 8.29. (a) One-dimensional compression and overconsolidation; (b) stiffness and overconsolidation (drawn for $\alpha = 1$).

8.11.1 Stiffness

If the stress σ'_z on the element in Fig. 8.28 changes, there will be corresponding vertical deformation described by the one-dimensional stiffness properties that we encountered in Section 4.5. The general relationship has the form:

$$\frac{E_o}{\sigma_{ref}} = \chi \left(\frac{\sigma'_z}{\sigma_{ref}}\right)^\alpha \quad (8.38)$$

where σ_{ref} is a reference stress introduced to leave dimensional consistency between the two sides of (8.38) and χ and α are soil parameters describing, respectively, the magnitude of the stiffness and the way in which it depends on stress level. For any change $\delta\sigma'_z$ in vertical effective stress, the corresponding vertical strain is $\delta\varepsilon'_z$:

$$\delta\varepsilon'_z = \frac{\delta\sigma'_z}{E_o} \quad (8.39)$$

We will see shortly why we need to distinguish this strain increment with the symbol ′.

In describing stiffness, we had to distinguish between normally compressed soils which were currently experiencing the maximum vertical stress they had ever experienced, and overconsolidated soils which had been more heavily loaded in the past (Section 4.7, Figs 8.14&8.29a). The overconsolidation ratio, $n = \sigma'_{z_{max}}/\sigma'_z$, describes the extent of this prior loading. The rules governing one-dimensional stiffness for normally compressed and overconsolidated soils were summarised in Section 4.7 (Fig. 8.29b):

1. If $\sigma'_z = \sigma'_{z_{max}}$ and $\delta\sigma'_z > 0$, then $n = 1$, $\alpha = \alpha_{nc}$ and $\chi = \chi_{nc}$ (normally compressed);
2. If $\delta\sigma'_z < 0$, then $\delta n > 0$, $n \geq 1$, and $\alpha = \alpha_{oc}$ and $\chi = \chi_{oc}$ (overconsolidated);
3. If $\sigma'_z < \sigma'_{z_{max}}$ and $\delta\sigma'_z > 0$, then $n > 1$, $\delta n < 0$, and $\alpha = \alpha_{oc}$ and $\chi = \chi_{oc}$ (overconsolidated);

8.11 Simple model of shearing ♣

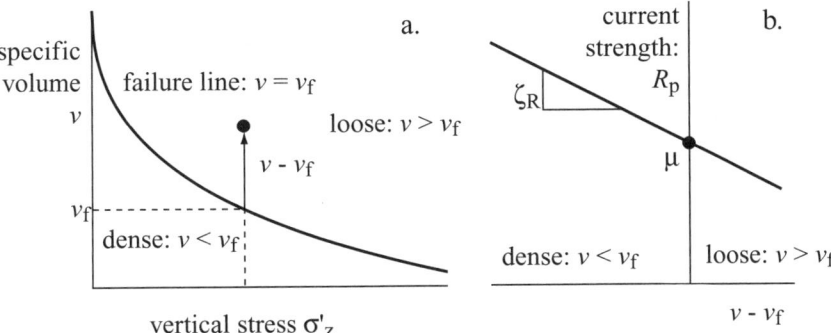

Figure 8.30. Failure line: (a) combinations of specific volume and normal stress; (b) current strength dependent on current specific volume.

where α_{nc} and α_{oc}, and χ_{nc} and χ_{oc} are the values of stiffness exponent and modulus number in (8.38) appropriate to the normally compressed and overconsolidated states, respectively.

8.11.2 Strength

As the shear stress τ on the element in Fig. 8.28 is increased, we expect eventually to reach the limit of the shear stress that can be supported – the strength of the soil. We have seen in Section 8.4 that the strength is dependent on density and is frictional in origin. We distinguish between the ultimate strength that is mobilised at large deformation and any temporary peak that may be seen on the way because the density is temporarily higher than it will be eventually. At large deformation the strength, expressed as the ratio of shear stress to normal stress $R = \tau/\sigma'_z$, has the value μ. If the density of the soil is presently different from the density appropriate to ultimate failure with the present normal stress, the available strength of the soil R_p is different from μ. We define the eventual relationship between specific volume and normal effective stress (Fig. 8.30a) by a simple relationship (compare (8.3)):

$$v_f = v_{min} + \Delta v \exp\left[-(\sigma'_z/\sigma_{ref})^\beta\right] \tag{8.40}$$

If the current specific volume is different from v_f, the current strength is R_p (Fig. 8.30b):

$$R_p = \mu + \zeta_R(v_f - v) \tag{8.41}$$

Thus, dense soils, with $v_f > v$, have current strength greater than the large deformation strength. Loose soils, with $v_f < v$, have current strength lower than the large deformation strength.

8.11.3 Mobilisation of strength

The ratio of shear stress to vertical effective stress, $R = \tau/\sigma'_z$, is a measure of the currently mobilised friction in our soil element: it is directly equivalent to the

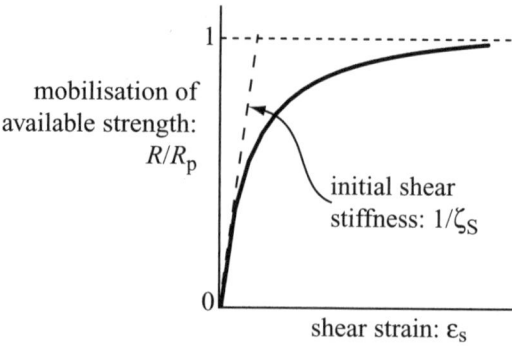

Figure 8.31. Hyperbolic mobilisation of strength.

ratio Q/P for the shear box (Fig. 8.6). The way in which strength is gradually mobilised as shear deformation increases is nonlinear, as we have seen in Fig. 8.7. Let us assume a simple hyperbolic relationship between mobilised friction R and shear strain ε_s which heads asymptotically towards the failure condition at large strains (Fig. 8.31):

$$\frac{R}{R_p} = \frac{\varepsilon_s}{\zeta_S + \varepsilon_s} \quad (8.42)$$

where ζ_S is a soil parameter which controls the initial shear stiffness of the element. Incrementally, this can be written:

$$\delta R = \frac{1}{R_p}\left[(R_p - R)^2 \frac{\delta \varepsilon_s}{\zeta_S} + R\delta R_p\right] \quad (8.43)$$

to remind us that the current strength R_p is not constant but depends on current density from (8.41).

8.11.4 Dilatancy

In presenting typical results of shear box tests, we observed the volume changes that occur in sands as they are sheared (Fig. 8.7) and related this dilatancy to the way in which granular materials were composed of rather rigid individual particles (Fig. 8.9). The phenomenon of dilatancy is a necessary part of the process by which a soil manages to move its density from its initial value to the value appropriate to the development of failure conditions under the current normal effective stress. Figure 8.7c shows an interpretation of the results in Figs 8.7a, b: there is a general correlation between the current slope of the volume change or vertical strain plots (Fig. 8.7b) and the current mobilised friction (Fig. 8.7a). The higher the mobilised friction, the more dramatic the rate of volume increase with continued shearing. In fact, a broad first order *stress-dilatancy* relationship could be proposed:

$$\frac{\delta z}{\delta x} = \zeta_D(\mu - R) \quad (8.44)$$

8.11 Simple model of shearing ♣

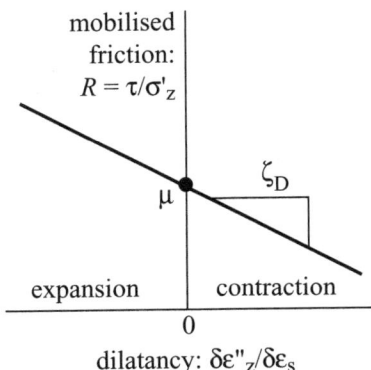

Figure 8.32. Shear box tests on sand: stress-dilatancy relationship.

which tells us that the ratio of normal movement to shear movement is proportional to the difference between the current mobilised strength, $R = \tau/\sigma'_z$, and the large deformation strength μ, with ζ_D being introduced as a soil parameter. For low mobilised strength, $R < \mu$, the soil compresses as it is sheared, $\delta z > 0$; for high mobilised strength, $R > \mu$, the soil expands or dilates as it is sheared, $\delta z < 0$.

We need to convert this relationship into a link between strain components for our soil element (Fig. 8.28), and we will write an exactly equivalent relationship (Fig. 8.32):

$$\frac{\delta\varepsilon''_z}{\delta\varepsilon_s} = \zeta_D(\mu - R) \tag{8.45}$$

where the symbol $''$ in $\delta\varepsilon''_z$ reminds us that this is a second, distinct, route to the generation of vertical strains. The vertical strain can change either through change in vertical effective stress, $\delta\varepsilon'_z$ (8.39), or through dilatancy, $\delta\varepsilon''_z$, (8.45), or both. The total vertical strain increment is the sum of these two components:

$$\delta\varepsilon_z = \delta\varepsilon'_z + \delta\varepsilon''_z \tag{8.46}$$

8.11.5 Complete stress:strain relationship

With a little manipulation, the various relationships can be written in incremental form and combined to deduce the stress increments $(\delta\sigma'_z, \delta\tau)$ that result from the application of any strain increments $(\delta\varepsilon_z, \delta\varepsilon_s)$:

$$\begin{aligned} \delta\sigma'_z &= E_o[\delta\varepsilon_z - \Psi_3\delta\varepsilon_s] \\ \delta\tau &= E_o[-\Psi_1\delta\varepsilon_z + (\Psi_2 + \Psi_1\Psi_3)\delta\varepsilon_s] \end{aligned} \tag{8.47}$$

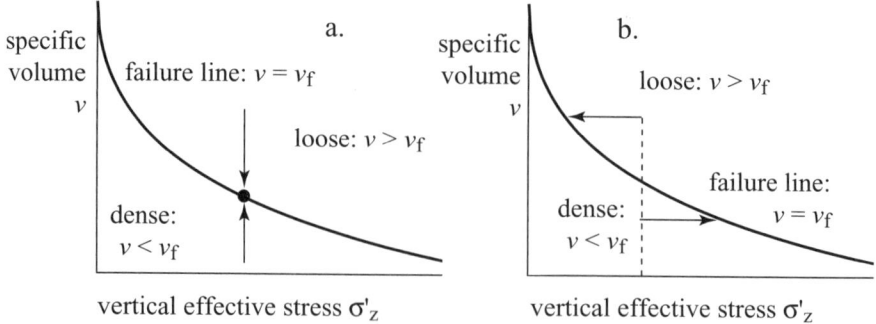

Figure 8.33. (a) Drained shearing with change of volume to failure; (b) undrained shearing with change in vertical effective stress to failure.

where:

$$\Psi_1 = \frac{\zeta_R R}{R_p} \left[\beta(v_f - v_{min}) \left(\frac{\sigma'_z}{\sigma_{ref}} \right)^\beta - \frac{v\sigma'_z}{E_o} \right] - R$$

$$\Psi_2 = \left[\frac{(R_p - R)^2}{\zeta_S R_p} + \frac{\zeta_R v R \Psi_3}{R_p} \right] \frac{\sigma'_z}{E_o} \qquad (8.48)$$

$$\Psi_3 = \zeta_D(\mu - R)$$

8.11.6 Drained and undrained response

The governing incremental equations (8.47) can be integrated numerically to generate the response to any particular loading or deformation history. Two obvious extremes to explore are the behaviour at constant vertical stress – the drained response, $\delta\sigma'_z = 0$ – and the behaviour when vertical deformation is prevented – the undrained response, $\delta\varepsilon_z = 0$.

With constant vertical stress, $\delta\varepsilon_z = \Psi_3 \delta\varepsilon_s$, and the vertical, volumetric strain is solely the result of dilatancy. The soil contracts if it is initially looser than the failure line $v > v_f$ or dilates if it is initially denser than the failure line $v < v_f$ (Fig. 8.33a). The shear stress:strain response is given by (Fig. 8.34):

$$\delta\tau = \Psi_2 E_o \delta\varepsilon_s \qquad (8.49)$$

Since this will in general imply volume change, we can think of this as a drained response of the soil. We observe that, no matter what the initial density or specific volume, the stress-strain response seeks out the large deformation strength μ, and the density changes, up or down, as required in order that the ultimate state of the soil should lie on the line of ultimate failure states (8.40).

Shearing at constant height, on the other hand, imposes a sort of conjugate mode of deformation on the soil. Whereas with constant vertical effective stress the height of the soil element will in general change (Fig 8.34), with constant height the vertical effective stress will in general change. We can understand this by studying the component parts of (8.46). Our imposed constraint controls the sum of the

8.11 Simple model of shearing ♣

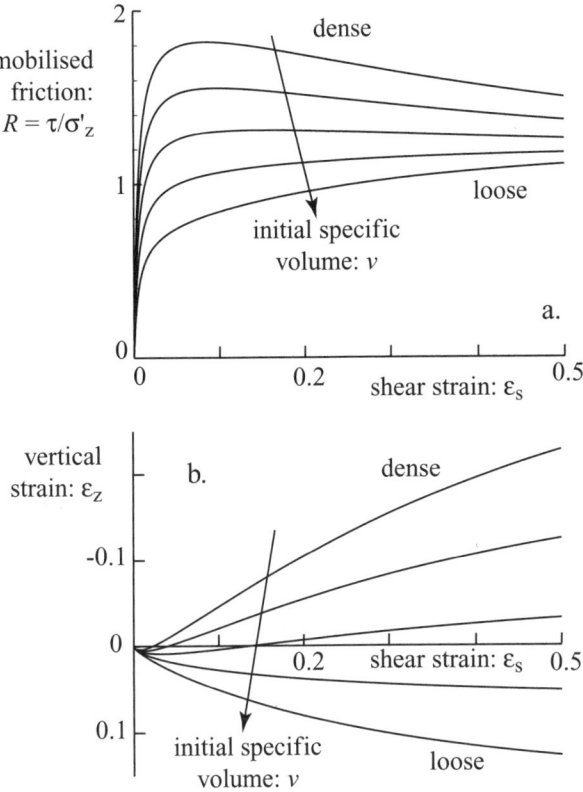

Figure 8.34. Drained shearing with different initial densities: (a) stress:strain response; (b) vertical strain.

two components not the individual components of vertical strain. The shearing produces dilatant vertical strains $\delta\varepsilon_z''$ and the vertical effective stress has to change to provide vertical strains $\delta\varepsilon_z' = \delta\sigma_z'/E_o = -\delta\varepsilon_z''$ to balance these, which would otherwise cause a change in height (volume) of the soil. Thus, if the soil is initially dense and is trying to expand, $\delta\varepsilon_z'' < 0$, the vertical effective stress has to increase to provide compensating compression, $\delta\varepsilon_z' > 0$ (Fig. 8.33b). Similarly, if the soil is initially loose and is trying to contract, $\delta\varepsilon_z'' > 0$, the vertical effective stress has to decrease to provide compensating expansion, $\delta\varepsilon_z' < 0$ (Fig. 8.33b). In either case, the total vertical strain increment is zero. The shear stress:strain response is given by (Fig. 8.35a):

$$\delta\tau = (\Psi_2 + \Psi_1\Psi_3)E_o\delta\varepsilon_s \tag{8.50}$$

and the effective stress path is given by (Fig. 8.35b):

$$\frac{\delta\sigma_z'}{\delta\tau} = -\frac{\Psi_3}{\Psi_2 + \Psi_1\Psi_3} \tag{8.51}$$

This constant volume shearing is equivalent to the undrained response of the soil. The change in vertical effective stress could occur by applying a constant external

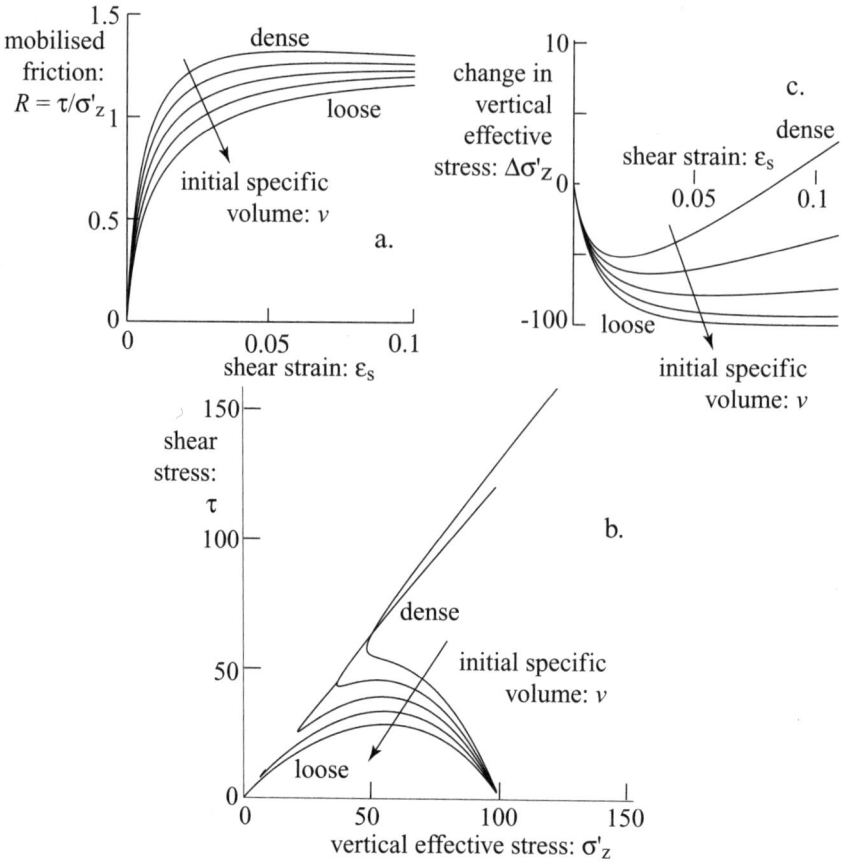

Figure 8.35. Undrained shearing with different initial densities: (a) stress:strain response; (b) effective stress path; (c) change in vertical effective stress.

vertical total stress $\delta\sigma_z = 0$, and then preventing any water from escaping from the saturated soil sample and allowing pore pressures to develop to produce the required changes in vertical effective stress: $\delta\sigma'_z = \delta\sigma_z - \delta u = -\delta u$. Alternatively, we could permit full drainage from the soil so that no pore pressures could develop, and then change the applied vertical stress to maintain the height of the sample constant. The changes in pore pressure that we discover with the first technique will exactly mirror the changes in external vertical stress that we see with the second (Fig. 8.35c). We can observe again that, no matter what the initial density or specific volume, the stress-strain response seeks out the large deformation strength μ, and the vertical effective stress changes, up or down as required, to force the ultimate state of the soil to lie on the line of ultimate failure states (8.40). In particular we might note that the effective stress path for the loosest (lowest density) sample in Fig. 8.35b ends up with more or less zero vertical effective stress and zero shear stress ($\sigma'_z = \tau \approx 0$): this corresponds to the condition of liquefaction which led to the rotation of the apartment blocks at Niigata in the 1964 earthquake (Fig. 1.10).

8.11.7 Model: summary

The current state of the soil is described by its specific volume v, and by the stresses σ'_z and τ, often combined in their ratio R. The quantities Ψ_1, Ψ_2 and Ψ_3 (8.48) are convenient short-hand groupings of the various constitutive properties of the soil together with the variables that define the current state of the soil. The model may appear a little complicated in its eventual formulation but it is in fact built up from a small number of simple building blocks.

- A one-dimensional stiffness relationship (8.39) defining normal stiffness E_o, which introduces soil properties χ, α and σ_{ref} (8.38) in order to reproduce the expected nonlinearity.
- A hyperbolic relationship between mobilised strength and shear strain (8.42) with soil property ζ_S.
- A stress-dilatancy relationship describing the volume changes accompanying shearing (8.45) requiring property ζ_D.
- A link between specific volume (or density) and vertical effective stress at ultimate failure, which introduces properties v_{min}, Δv and β (8.40).
- A link between current strength and volumetric distance from this failure line (8.41), requiring properties μ and ζ_R.

Each of these relationships is mathematically rather simple: the apparent complexity arises because of their interlocking.

However, the model is rather powerful. It shows how the shearing response depends on the initial density of the soil. It shows how the volume changes accompanying shearing inexorably move the soil from its initial condition to the failure line in drained tests, and the vertical effective stress has to change to allow the soil to move to the failure line in undrained tests.

For any complex model, there is a simpler model lurking inside which we can find by eliminating some of the special effects. Thus, we could eliminate the nonlinearity of normal stiffness by setting $\alpha = 0$; we could eliminate the dilatancy (the change in volume induced by shearing) by setting $\zeta_D = 0$; we could eliminate the variation of strength with density by setting $\zeta_R = 0$. Exploration of the consequences of switching off these features will be left for the reader to pursue.

8.12 Summary

Here is a concise list of the key messages from this chapter, which are also encapsulated in the mind map (Fig. 8.36).

1. In many geotechnical systems, failure occurs by sliding on a clear slip surface, mobilising the shear strength of the soil.
2. The shear box is a laboratory test device which mimics the development of a failure surface in soil.
3. A simple model proposes a frictional strength linking maximum shear stress with the normal effective stress on the surface of failure.

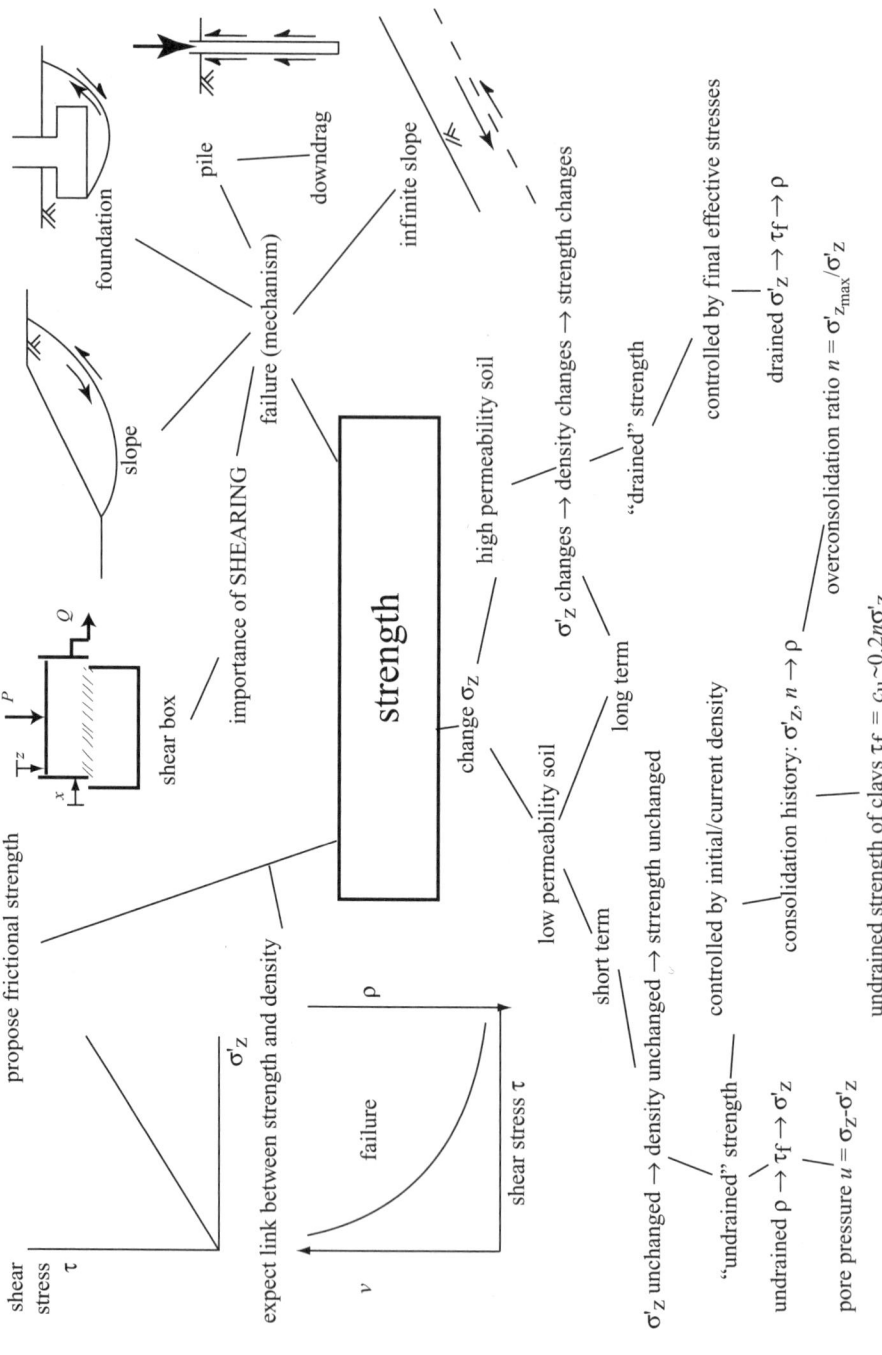

Figure 8.36 Mind map: strength

4. We also expect the strength to be linked with the density of the soil – the denser the soil, the higher the strength.
5. This simple model allows us to make clear statements about drained and undrained strengths of soils.
6. High permeability soils (or low permeability soils after a long time) are able to change in volume when the normal stress is changed and the strength and the final density are controlled by the final effective stress.
7. Low permeability soils are not able to change in volume rapidly when the normal stress is changed: the strength is controlled by the initial density. The strength then implies a certain normal effective stress at failure, and the difference between the total and effective normal stresses allows us to estimate the pore pressure at failure.
8. A model can be generated to describe the shearing and compression of soils and to reproduce the effects of different initial densities and different drainage conditions.

8.13 Exercises: Strength

1. A sample of normally compressed clay whose strength is described by an expression of the form $\tau_f = \Lambda \sigma'_z$ with $\Lambda = 0.3$, has current vertical total stress $\sigma_z = \sigma'_{z_{max}} = 170$ kPa and pore pressure $u = 50$ kPa. What is the current vertical effective stress? What is the current strength?
2. The total stress on the sample in Question 1 changes to 250 kPa without drainage (so that there is no change in density of the sample). What are the new values of pore pressure, vertical effective stress and strength?
3. The total stress on the sample of Question 2 is maintained constant at 250 kPa and the pore pressure falls slowly to zero: drainage is permitted and the density of the clay is able to change. What is the eventual value of the vertical effective stress? Does the density of the clay increase or decrease with time? What is the eventual undrained strength?
4. A shallow slope failure occurred at Jackfield, Shropshire, after a period of heavy rain in the winter 1952–1953.[9] The slope angle was found to be about 10.5° and failure was found to have occurred on a thin surface parallel to the slope at a depth of about 5 m. What shear strength must have been mobilised on the failure plane? The unit weight of the clay was 20.4 kN/m^3.

   Laboratory tests showed an angle of shearing resistance for the clay of about 21°. Field tests indicated that seepage was occurring more or less parallel to the slope. Calculate the mobilised angle of friction on the failure surface and confirm that the margin of stability would have been small.
5. Figure 8.37 shows an approximate section through the landslide that occurred at Po Shan Road in Hong Kong in 1972, shown in Fig. 5.2. A slip surface formed at

---

[9] Henkel, D.J. & Skempton, A.W. (1955) A landslide at Jackfield, Shropshire in a heavily overconsolidated clay. *Géotechnique* **5** 2, 131–7.

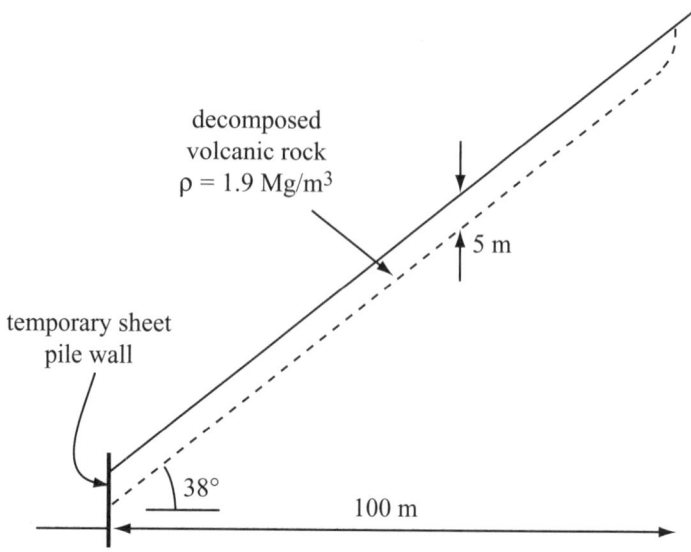

Figure 8.37. Schematic section through landslide in Fig. 5.2 for Question 5.

a depth of about 5 m in a natural slope (with angle $\theta \approx 38°$) composed of decomposed volcanic rock with density $\rho = 1.9$ Mg/m^3. The sliding material extended a horizontal distance of about 100 m from a temporary excavation cut into the slope in which support to the slope above was provided by a sheet pile wall. In Hong Kong, the climatic conditions are such that the soils may sustain significant negative pore pressures for much of the year, but when it rains heavily the rain water infiltrates the ground and removes these negative pore pressures – just like the sand castle in Fig. 8.25c.

The soil has an angle of friction $\phi = 30°$. What negative pore pressure is required to just guarantee stability of the slope in the absence of any additional support?

After extremely heavy rainfall, a seepage regime in the soil parallel to the slope is rapidly created ($\psi = \theta$). What angle of friction would be required to ensure stability of the slope? Given that the available friction is only $\phi = 30°$, what force parallel to the slope would the sheet pile wall need to provide to support the layer of soil that is on the point of sliding? [Hint: Assume that the stresses are uniform along the slip surface and write equations of equilibrium parallel to and orthogonal to the slope.]

6. At the moment that the fall-cone of Fig. 8.27a is released, the cone of mass m experiences a downward gravitational attraction mg. Newton's Law of Motion tells us that the cone will accelerate with acceleration $a = g$ and begin penetrating the clay. In general, when the penetration of the cone is z the upwards resistance to continued penetration is $c_u z^2 / k_\omega$, and the equation of motion tells us that:

$$a = v \frac{dv}{dz} = g - \frac{c_u z^2}{k_\omega m} \tag{8.52}$$

8.13 Exercises: Strength

The penetration of the cone in Fig. 8.27 is a dynamic process leading to an eventual penetration d. At what penetration d_s would a cone of the same tip angle ω encounter a resistance to penetration equal to its weight mg if it were to be pushed in at essentially zero velocity? [Hint: The fall-cone will have zero acceleration at the moment when its penetration is $z = d_s$ but it still has a non-zero velocity, and eventually comes to rest with $z = d > d_s$. You are asked to discover the ratio d_s/d.]

7. This exercise brings together the stiffness model of Chapter 4 and the strength model of the present chapter.

   The one-dimensional stiffness of a sandy-silt is characterised by the stiffness properties $\alpha_{nc} = \alpha_{oc} = 0.5$, $\chi_{nc} = 100$, $\chi_{oc} = 400$. When the vertical effective stress $\sigma'_z = 1$ kPa, $v = 1.9$. By appropriate integration of (4.21) find an expression for the normal compression line linking values of σ'_z and v.

   The failure conditions for the sandy-silt are described by (8.3) and (8.2) with $\phi = 25°$, $v_{min} = 1.2$, $\Delta v = 1.5$, $\beta = 0.3$, $\sigma_{ref} = 100$ kPa.

   a. The soil is normally compressed to a vertical effective stress $\sigma'_z = 100$ kPa. What is the specific volume? What are the drained and undrained strengths? What is the vertical effective stress at undrained failure?

   b. The soil is further compressed to a vertical effective stress $\sigma'_z = 500$ kPa. What is the specific volume? What are the drained and undrained strengths? What is the vertical effective stress at undrained failure?

   c. The soil is unloaded to a vertical effective stress $\sigma'_z = 100$ kPa. What is the specific volume? What are the drained and undrained strengths? What is the vertical effective stress at undrained failure?

   d. The additional properties of the soil required to complete the deformation model are $\zeta_S = 0.005$, $\zeta_D = 1$, $\zeta_R = 0.3$, $\mu = \tan\phi$. Estimate the vertical strain ε_z and shear strain ε_s when the stress ratio $R = \mu/2$ for each of the samples in (a), (b) and (c). [Hint: The numerical integration of (8.47) can be achieved with a small number of increments.]

9 Soil-structure interaction

9.1 Introduction

Soil-structure interaction is one of those interface topics which cannot be treated successfully either as a purely structural problem or as a purely geotechnical problem. A holistic approach is required to the modelling – the identification of the essential details of the problem – and to the subsequent analysis. The geotechnical system in this case is the sum of all the geotechnical and structural elements, and the response of the system will certainly depend on some combination of properties of both the soil and the structure. If the ground and the structure are both behaving elastically, then simple configurations lead to exact analyses. While it has to be admitted that the problems that can be analysed are somewhat idealised, there is sufficient realism to demonstrate and support the important messages of soil-structure interaction.

Let us start with a thought experiment that will seem quite remote from soil-structure interaction. Suppose that we have a quarter kilogram (or half pound) packet of butter (unwrapped) on a plate. We also have a penknife or some other knife with a short, stiff blade, and a palette knife with a rather flexible blade. We place the flat side of the blades of the knives on the block of butter in turn and try to make an impression in the surface. The short, stiff blade will penetrate without difficulty (Fig. 9.1a); the palette knife blade will just bend (Fig. 9.1b). If we were to repeat the experiment with a bowl of blancmange or jelly or thick soup (minestrone?), then the flexible blade would have no difficulty in penetrating the surface. However, if we were to repeat the experiment using a piece of very hard cheese (parmigiano?) then even the stiff blade would have difficulty making an impression.

We have used terms like "short" and "stiff" and "flexible" and we know what they mean, but really they are describing relative properties of one blade to another. In terms of the effect that the blades have on the butter, soup and cheese we cannot predict the extent or ease of penetration without also knowing something about the material into which the blade is penetrating. It seems that we cannot make a definitive statement unless we somehow define "stiff" and "flexible" in relation to the stiffness (or strength) properties of the butter, soup or cheese.

9.1 Introduction

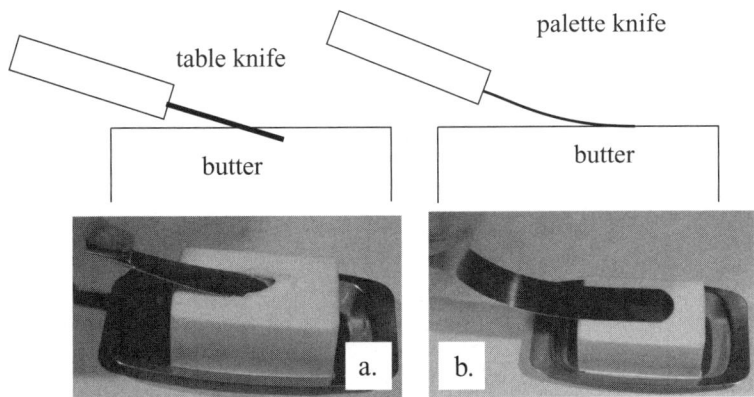

Figure 9.1. Penetration of block of butter with (a) short, stiff knife blade; (b) flexible palette knife blade.

Now move up in scale and imagine that we are trying to support the load from a column in a building on a foundation. If the ground conditions are good enough, then there will be some economic advantage in using a spread foundation – of area considerably larger than the column – made of reinforced concrete to transfer the load to the ground at a shallow depth (Fig. 9.2). The stiffer the foundation, the more reinforcement that will be needed and the more expensive the foundation. However, by analogy with our knife blades and the block of butter, we can expect that there will be a trade-off between the flexibility of the foundation and the amount of relative deformation or bending that the foundation will experience. A thin, lightly reinforced flexible foundation will tend to settle most directly under the load from the column, whereas the ends may hardly move at all (Fig. 9.2a) – recall the palette knife. A very stiff, heavily reinforced foundation will settle uniformly as a rather rigid object (Fig. 9.2b) – recall the short, stiff knife blade.

Schematic diagrams of foundation settlement are shown in Fig. 9.3 for a range of foundation flexibilities ranging from absolutely rigid (Fig. 9.3a), settling by an identical amount across the whole width of the foundation, to absolutely flexible (Fig. 9.3e), which settles only directly under the applied load (imagine pushing a thin sheet of plastic into the block of butter using the thin edge of the blade of

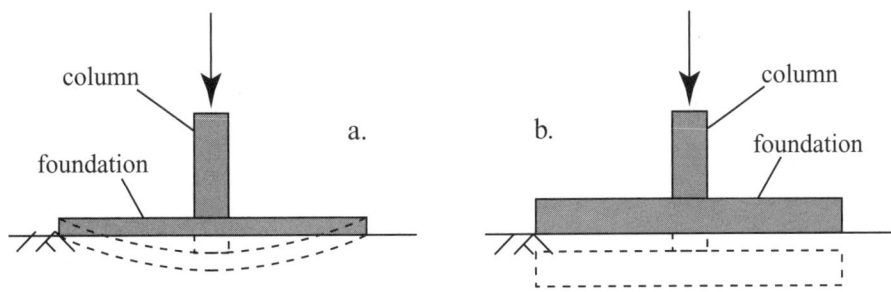

Figure 9.2. Column load supported on (a) flexible spread foundation and (b) stiff spread foundation.

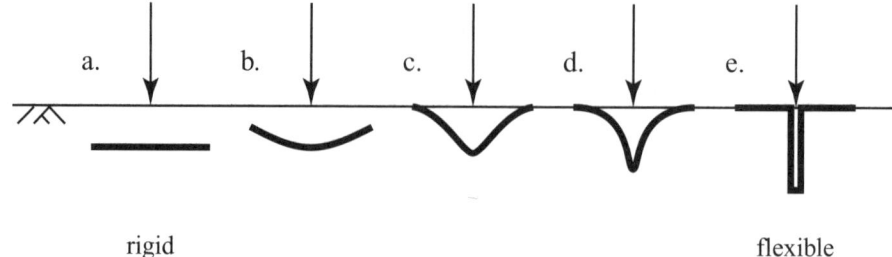

Figure 9.3. Influence of foundation stiffness on settlement across foundation.

a knife). These are extremes, but we expect again that the detail of this response cannot only depend on the structural properties of the concrete foundation slab but must also be affected by the properties of the ground. What will constitute a flexible or a stiff foundation in the character of its performance under load cannot actually be assessed without knowing something also about the stiffness properties of the ground on which the foundation is placed. This is the problem analysed in Section 9.4.

Next, imagine that you are pitching a tent and need to anchor the fly-sheet to ensure that it does not blow away (Fig. 9.4). The ties are held down using tent pegs. We might use rather stiff chunky wooden pegs (Fig. 9.4b) or we might use much more flexible metal pegs (Fig. 9.4c). If the wind gets up, or the tension in the ties is too great, or the ground is too soft, then the peg may move or deform. The stiff peg will merely rotate (Fig. 9.4b) but the flexible metal peg will probably bend (Fig. 9.4c). However, if the ground is soft enough then even the flexible peg will only rotate without bending. The performance of the pegs depends on the stiffness properties of both the peg and the ground.

Replace the tent with a floating offshore structure for oil exploration and production which has to be tethered to the seabed: the scale is different but the mechanics are the same. The tent pegs are replaced by piles which are pulled sideways by the tethering cables. The response of these piles will depend on the stiffness properties of the piles themselves and of the ground into which they have been driven. This is the problem analysed in Section 9.5.

Both these examples (spread foundations and laterally loaded piles) involve bending of the structural elements, and the underpinning analysis of bending of beams is summarised in Section 9.3. However, we begin with a slightly simpler

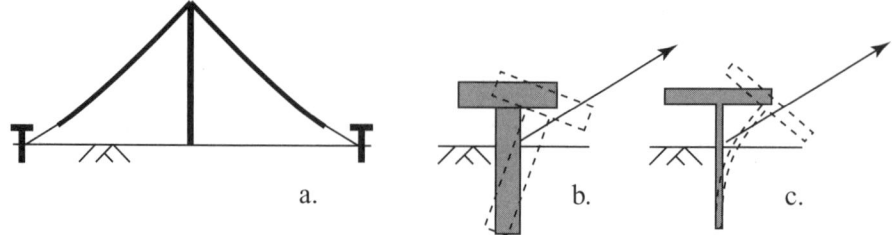

Figure 9.4. (a) Fly-sheet of tent anchored by tent pegs: (b) stiff peg; (c) flexible peg.

9.2 Pile under axial loading ♣

example which involves only axial loading of the structural element: an axially loaded pile.

9.2 Pile under axial loading ♣

As we have seen in Sections 1.3 and 8.8, a pile is a structural element that is used to transfer a load (such as the column load in Fig. 9.2) into the ground at a depth below the surface. The load is transferred to the ground by a combination of shaft resistance down the sides of the pile and end bearing at its toe. In Section 8.8 we were considering the strength of the soil at the interface with the pile and therefore really only concerned with failure conditions for this interface. Now we want to explore the possibilities of making some estimates of the performance of such a pile under working conditions in which we can treat both the pile and the ground as elastic. We will model the interface between the pile and the soil in a somewhat simplistic way in order that we may be able to reduce the problem, which is obviously really three dimensional, to a one-dimensional problem in which the single dimension is the position down the pile.

To appreciate at least qualitatively the nature of the soil-structure interaction in this problem, let us initially consider two extremes. In the first, the pile is extremely stiff axially and the soil is very weak. Because the pile is stiff, the axial displacement will be virtually the same all the way down the pile, and because the soil is weak, there will be very little load shed to the ground through shaft resistance along the length of the pile. Just about all of the load applied to the pile at the ground surface will be transferred to the toe of the pile to be carried in end bearing.

At the other extreme, we will have an extremely compressible pile in very stiff soil. As any point on the pile moves downward relative to the distant undeforming ground, the stiffness of the soil will lead to the generation of shear stresses round the pile, and this shaft resistance will reduce the load in the pile. As we go deeper into the ground, the load shed to the ground increases but the deformation of the pile reduces because of the falling axial load. The settlement at the toe of the pile will be much less than that at the ground surface – in fact, if the pile is compressible enough, then it may be that there is no load left in the pile by the time we get to its toe, and all the load will have been transferred to shaft resistance. We have again used terms such as "stiff", "weak", "compressible" to describe the pile and the soil but we will not be surprised to find that they are in fact relative terms which have to combine properties of both the soil and the structural elements.

We will analyse an elastic pile which is being loaded axially in an elastic soil. We can obtain an exact closed-form solution for the load distribution within the pile and the settlement distribution down the pile. We will assume the pile to be circular in cross-section for the calculation of its perimeter and cross-sectional area.

At a depth z down the pile (Fig. 9.5), the pile has a settlement ς relative to the distant undeforming ground. The shear stress τ that develops at the interface between the pile and the soil will be dependent on the stiffness of the soil. Since we are assuming that the soil is elastic, we can plausibly assume that this shear stress τ

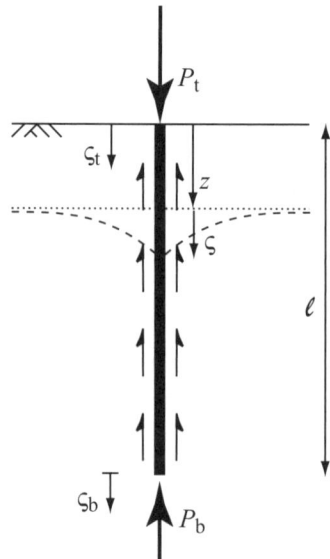

Figure 9.5. Axially loaded pile in elastic soil.

is linearly related to the settlement ς normalised with the radius r_o of the pile.[1]:

$$\tau = \lambda \frac{\varsigma}{r_o} \qquad (9.1)$$

Now we can consider the equilibrium of an element of the pile of thickness δz at depth z (Fig. 9.6). The axial load at the top of the element is P and at the bottom of the element $P + \delta P$. The settlement of the pile relative to the distant soil is ς and, as a result of this settlement, there is a shear stress τ round the perimeter of the pile which will support part of the axial load. Equilibrium tells us that:

$$\delta P = -2\pi r_o \delta z \tau = -2\pi \lambda \varsigma \delta z \qquad (9.2)$$

or in the limit:

$$\frac{dP}{dz} = -2\pi \lambda \varsigma \qquad (9.3)$$

and, as expected, δP turns out to be negative.

The pile is elastic and compresses under axial stress. Compression of the pile implies that the settlement varies down the pile. The settlement at the top of the element is ς and at the bottom $\varsigma + \delta \varsigma$ (Fig. 9.6). If the settlement varies across the element, then there will be a tensile axial strain in the pile $\varepsilon_a = \delta \varsigma / \delta z$, and an axial

---

[1] A more extensive discussion of this axial pile analysis is provided by Fleming, W.G.K., Weltman, A.J., Randolph, M.F. and Elson, W.K. (1985) *Piling engineering*. Surrey University Press, Glasgow and John Wiley, New York. They, unconstrained by the need to consider only a single dimension, link the shear stress τ with the shear modulus G of the soil and find that $\tau \approx \varsigma G/4r_o$. Thus, the stiffness constant used here is $\lambda = G/4$.

9.2 Pile under axial loading ♣

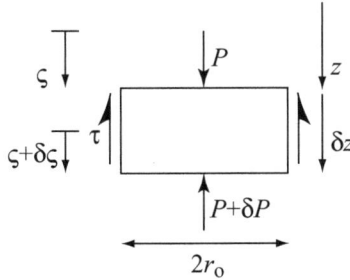

Figure 9.6. Equilibrium of element of axially loaded pile in elastic soil. Note: exaggerated indication of settlement.

stress related to the axial strain through Young's modulus E_p for the pile:

$$\frac{d\varsigma}{dz} = -\frac{P}{\pi r_o^2 E_p} \tag{9.4}$$

The negative sign is required because the pile is in compression and every element of the pile must be getting shorter, $d\varsigma/dz < 0$.

Combining (9.3) and (9.4), the governing equation for the axially loaded pile is:

$$\frac{d^2\varsigma}{dz^2} = \frac{2\lambda\varsigma}{E_p r_o^2} \tag{9.5}$$

This has a general solution:

$$\varsigma = J_1 \exp[\Theta z] + J_2 \exp[-\Theta z] \tag{9.6}$$

where:

$$\Theta = \frac{1}{r_o}\sqrt{\frac{2\lambda}{E_p}} \tag{9.7}$$

The problem is characterised by the value of the parameter Θ which has dimensions of length^{-1} and combines, as expected, the stiffness of the pile E_p and the stiffness of the pile-soil interaction λ. We have to apply the general solution (9.6) to the boundary conditions of our particular problem.

For all piles, the load at the top of the pile is equal to the applied load P_t:

$$P_{z=0} = -\pi r_o^2 E_p \left(\frac{d\varsigma}{dz}\right)_{z=0} = P_t \tag{9.8}$$

where the negative sign is required because the pile is in compression and the relative pile-soil movement ς decreases down the pile as load is shed to the surrounding soil.

If the pile is very long (where "long" is a relative term) then as $z \to \infty$, $\varsigma \to 0$. Then, in (9.6), $J_1 = 0$ and $J_2 = (P_t/\pi r_o)/\sqrt{(2\lambda E_p)}$ and the load varies exponentially down the pile:

$$\frac{P}{P_t} = \exp[-\Theta z] \tag{9.9}$$

or:

$$\ln \frac{P}{P_t} = -\Theta z \qquad (9.10)$$

We could introduce an effective length ℓ_{100} at which the axial load in the pile had dropped to 1% of its top value, $P/P_t = 0.01$. We have $\Theta \ell_{100} = \ln 100$ or $\ell_{100}/r_o = \ln 100 \sqrt{E_p/2\lambda}$ and the definition of "long" involves E_p and λ as well as the pile radius r_o. If the pile is very compressible (low E_p) and the soil stiff (high λ), then very little load reaches the base of the pile – in accord with our qualitative introductory discussion.

On the other hand, for a pile with finite length ℓ (Fig. 9.5), there will in general be some load P_b remaining at the base of the pile which must be compatible with the base settlement ς_b of the pile and this is the boundary condition at $z = \ell$. Assuming elastic base load transfer (closed form analysis is not straightforward without such an assumption), then, for $z = \ell$:

$$\frac{P_b}{\pi r_b^2} = \lambda_b \frac{\varsigma_b}{r_b} \qquad (9.11)$$

where r_b is the radius of the base of the pile (which might be deliberately constructed by a technique called "under-reaming" to a larger radius than the pile itself).[2] The imposition of this boundary condition in (9.6) leads to a more cumbersome result:

$$J_1 = \frac{P_t}{2\pi r_o^2 E_p \Theta} \frac{(\xi - 1)(1 - \tanh \Theta \ell)}{1 + \xi \tanh \Theta \ell} \qquad (9.12)$$

$$J_2 = \frac{P_t}{2\pi r_o^2 E_p \Theta} \frac{(\xi + 1)(1 + \tanh \Theta \ell)}{1 + \xi \tanh \Theta \ell} \qquad (9.13)$$

where:

$$\xi = \frac{r_o^2 E_p \Theta}{\lambda_b r_b} = \frac{r_o}{r_b} \frac{\sqrt{2 E_p \lambda}}{\lambda_b} \qquad (9.14)$$

and ξ combines the geometrical ratio r_o/r_b with a composite stiffness ratio which includes not only the pile stiffness E_p but also both the pile-soil transfer stiffnesses λ and λ_b.

The settlement ς_t at the top of the pile, at $z = 0$ (Fig. 9.5), which is required to estimate the overall pile stiffness P_t/ς_t, is:

$$\varsigma_t = \frac{P_t}{\pi r_o^2 E_p \Theta} \frac{(\xi + \tanh \Theta \ell)}{(1 + \xi \tanh \Theta \ell)} \qquad (9.15)$$

and the load variation down the pile is given by:

$$\frac{P}{P_t} = \frac{(\xi + 1)(1 + \tanh \Theta \ell) e^{-\Theta z} - (\xi - 1)(1 - \tanh \Theta \ell) e^{\Theta z}}{2(1 + \xi \tanh \Theta \ell)} \qquad (9.16)$$

---

[2] Again, pushing at the boundaries of our one-dimensional constraints, the base stiffness could be related to elastic properties (Poisson's ratio ν_b and shear modulus G_b) of the continuous soil beneath the base of the pile, $\lambda_b = 4G_b/[\pi(1 - \nu_b)]$, which is the equation for the stiffness of a circular plate on the surface of an elastic material.

9.2 Pile under axial loading ♣

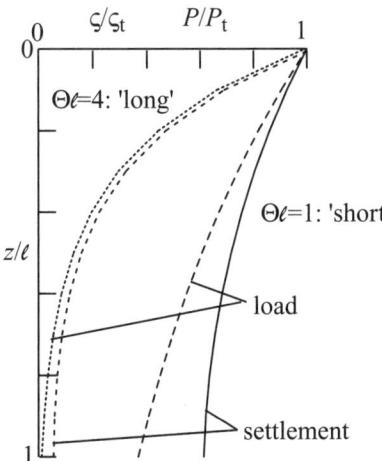

Figure 9.7. Axially loaded pile in elastic soil; distributions of axial load and axial displacement.

The parameter Θ (9.7) that characterises the solution is a function of the ratio of the stiffnesses of the pile and the soil shaft resistance transfer. In fact, Θ always appears in the solution in combination with either the length or a distance down the pile. Thus $\Theta\ell = (\ell/r_o)\sqrt{2\lambda/E_p}$ is the appropriate dimensionless group which controls the overall response of a pile of length ℓ so that $\Theta z = \Theta\ell(z/\ell)$. We have managed to characterise the entire performance of the pile (9.16) in terms of dimensionless groups for load and position P/P_t and z/ℓ, and for appropriate combinations of material and geometric properties $\Theta\ell$ and ξ. The result is thus of completely general applicability. Results are shown in Fig. 9.7 for $\Theta\ell = 1$ and 4 – these values produce classic "short" and "long" pile response, respectively. For a short pile (low values of $\Theta\ell$), the base of the pile takes a significant load, and develops a correspondingly significant settlement. For a longer pile, little load reaches the base of the pile.

9.2.1 Examples

1. Let us suppose that we have a concrete pile of radius $r_o = 0.25$ m and Young's modulus $E_p = 25$ GPa. The pile is installed in a firm clay with undrained strength $c_u = 200$ kPa. The stiffness of clays is quite well correlated with undrained strength as we have seen – both depend on the density of packing and loading history. Assume that the stiffness of the pile-soil load transfer is $\lambda = 50c_u$. Calculate the distance ℓ_{100} down the pile at which the load has fallen to 1% of the surface value.
First, calculate $E_p/\lambda = 2500$. For the load to fall to 1% of the surface value, $\ell_{100}/r_o = \ln 100\sqrt{(E_p/2\lambda)} = 163$, corresponding to an effective length of $\ell_{100} \approx 41$ m.

2. A pile of length $\ell = 25$ m and radius $r_o = r_b = 0.3$ m is made of concrete with Young's modulus $E_p = 25$ GPa in soil with shaft shear transfer stiffness $\lambda =$

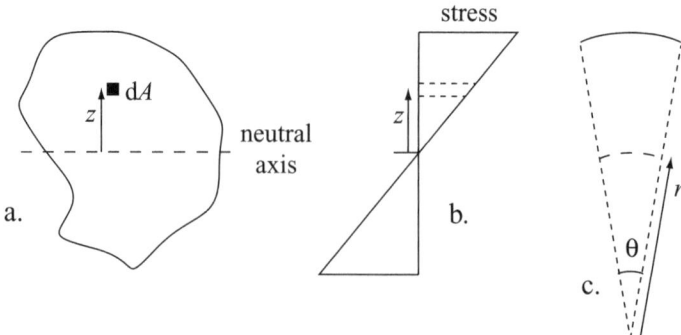

Figure 9.8. Bending of beam.

6.25 MPa and base load transfer stiffness $\lambda_b = 125/\pi$ MPa. Find the stiffness of the pile P_t/ς_t.

First, calculate the various stiffness ratios: $\xi = 14.05$, $\Theta\ell = 1.863$, $\tanh\Theta\ell = 0.953$. Then from (9.15), the stiffness of the pile is $P_t/\varsigma_t = 505$ MN/m. The proportion of the surface settlement seen at the base of the pile is 28% but the proportion of the surface load received at the base of the pile is only 2%. It is seen in Fig. 9.7 that the axial load in the pile typically falls off with depth faster than the pile settlement.

9.3 Bending of an elastic beam ♣

The soil-structure interaction problems that we will analyse in Sections 9.4 and 9.5 involve the bending of beam elements. The problems remain one-dimensional – the one dimension being the distance measured along the beam. The basic building block is the bending equation, the differential equation which links the loading on an elastic beam with the resulting deformations. This will almost certainly have been encountered in parallel courses on structural mechanics but is included here for completeness and to emphasise that soil-structure interaction cannot be understood without understanding the basic responses of both structural and geotechnical elements.

We consider bending about one axis (Fig. 9.8). The analysis of the bending of a beam of arbitrary cross-section assumes that plane sections remain plane, which is to say that a straight line across the section of the beam remains straight when the beam is bent under external loading so that the deformation can be completely described by the radius of curvature r (Fig. 9.8c). If we think about "fibres" in the section of the beam – such as the element dA shown in Fig. 9.8a – we can imagine that, as a result of the bending, some fibres will be extended and develop tensile stresses and others will become shorter and develop compressive stresses. There will be a line across the section – the *neutral axis* (Fig. 9.8) – separating the parts of the beam which are in tension and those that are in compression. Along the neutral axis, the

9.3 Bending of an elastic beam ♣

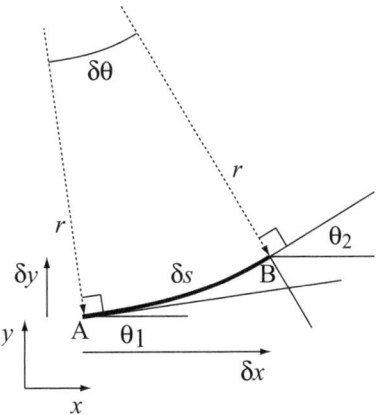

Figure 9.9. Curvature.

fibres of the beam are neither getting longer nor getting shorter and develop neither tensile nor compressive stresses.

The beam is bending with a radius of curvature r, measured to the neutral axis; r is large in comparison with the dimensions of the cross-section of the beam (the geometry in Fig. 9.8c is deliberately exaggerated). If we know the strain developing parallel to the axis of the beam at any point in the section as a result of the bending, then we can calculate the corresponding axial stress from the Young's modulus E of the material of the beam. The neutral axis is by definition not straining, so for a section of the beam subtending an angle θ at the centre of curvature, the unstretched length of any fibre in the section (at the neutral axis) is $r\theta$. The stretched length of a fibre at distance z from the neutral axis is $(r+z)\theta$, and hence the tensile strain is z/r and the stress Ez/r. The contribution to the moment provided by an elemental area δA at a distance z from the neutral axis is then $zE(z/r)\delta A$, and the moment provided by the complete section is the integral over the entire cross section A of the elemental contributions:

$$M = \int_A \frac{E}{r} z^2 \mathrm{d}A = \frac{EI}{r} \tag{9.17}$$

where:

$$I = \int_A z^2 \mathrm{d}A \tag{9.18}$$

is termed the second moment of area for the section of the beam. The combination EI is known as the flexural rigidity of the beam section.

Next, we need to relate curvature or radius of curvature to the differential geometry of the deflected beam. Figure 9.9 shows a segment AB of a curved beam. The segment has length δs and subtends an angle $\delta\theta$ at the centre of curvature of the segment. The radius of curvature r is then:

$$r = \frac{\delta s}{\delta\theta} \tag{9.19}$$

and the inverse of the radius is the curvature κ:

$$\kappa = \frac{1}{r} = \frac{\delta\theta}{\delta s} \tag{9.20}$$

The length of the segment δs is given approximately by:

$$\delta s = \sqrt{(\delta x^2 + \delta y^2)} \tag{9.21}$$

The slope angle θ_1 at A is related to the slope of the curve:

$$\tan\theta_1 = \frac{dy}{dx} \tag{9.22}$$

and the slope angle θ_2 at B:

$$\tan\theta_2 = \frac{dy}{dx} + \frac{d}{dx}\left(\frac{dy}{dx}\right)\delta x = \frac{dy}{dx} + \frac{d^2y}{dx^2}\delta x \tag{9.23}$$

The subtended angle $\delta\theta$ is the difference between the slope angles at the two ends of the segment:

$$\delta\theta = \theta_2 - \theta_1 \tag{9.24}$$

so that:

$$\tan\delta\theta = \tan(\theta_2 - \theta_1) = \frac{\tan\theta_2 - \tan\theta_1}{1 + \tan\theta_2 \tan\theta_1}$$

$$= \frac{\frac{d^2y}{dx^2}\delta x}{1 + \left(\frac{dy}{dx}\right)^2} \tag{9.25}$$

neglecting third order small quantities in the denominator.

For an infinitesimal length of beam we can write $\tan\delta\theta \approx \delta\theta$, and then combining (9.25), (9.20) and (9.21) we find an expression for the curvature:

$$\kappa = \frac{\frac{d^2y}{dx^2}}{\left[1 + \left(\frac{dy}{dx}\right)^2\right]^{3/2}} \tag{9.26}$$

For a beam undergoing small deformations, we can write the curvature approximately and simply as:

$$\kappa \approx \frac{d^2y}{dx^2} \tag{9.27}$$

since $(dy/dx)^2 \ll 1$. Then combination with (9.17) shows us that:

$$M = EI\frac{d^2y}{dx^2} \tag{9.28}$$

The final step in generating the differential equation governing the mechanical behaviour of a beam in bending is shown in Fig. 9.10. An infinitesimal section of beam of length δx is loaded with a transverse force w per unit length. Imagining the section cut from a longer beam, there will be force resultants on each end of

9.3 Bending of an elastic beam ♣

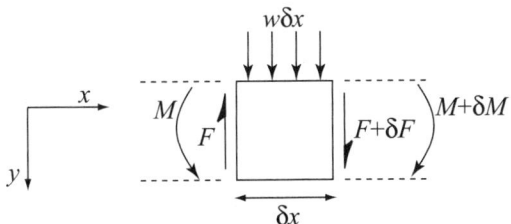

Figure 9.10. Element of beam.

the section: a shear force F and a bending moment M. In general, these will change with position along the beam.

We have to declare and take care over our sign convention. Figure 9.10 shows a reference set of (x, y) axes with y, the transverse deflection of the beam, measured positive downwards. From (9.27), positive curvature requires a slope dy/dx which increases with position x. We define a positive moment as one which tends to encourage such a downward curvature – known as a *hogging* moment, with tension in the upper fibres of the beam. We define a positive shear force as a *clockwise* shear force, with the beam beyond the cut tending to pull down on the neighbouring section of the beam. The directions of the moments and shear forces shown in Fig. 9.10 are consistent with these definitions. The direction of the applied loading w is positive in the direction y of deflection of the beam.

Vertical (transverse) equilibrium of the section tells us:

$$\delta F + w\delta x = 0 \tag{9.29}$$

and moment equilibrium of the section tells us:

$$\delta M + F\delta x = 0 \tag{9.30}$$

neglecting second order small quantities. Therefore, with (9.28):

$$F = -\frac{dM}{dx} = -EI\frac{d^3 y}{dx^3} \tag{9.31}$$

Combining these:

$$w = -\frac{dF}{dx} = \frac{d^2 M}{dx^2} \tag{9.32}$$

and, again substituting for the moment from (9.28):

$$w = EI\frac{d^4 y}{dx^4} \tag{9.33}$$

and this is the equation that we will use in our next examples of analytical soil-structure interaction.

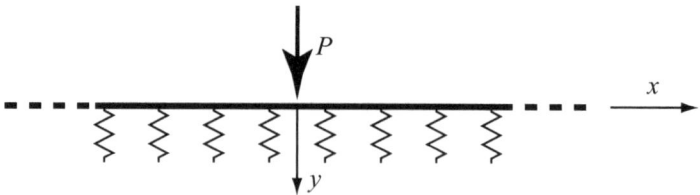

Figure 9.11. Beam on elastic foundation.

9.4 Elastic beam on elastic foundation ♣

Geotechnical engineering regularly involves the design of shallow, near-surface foundations (Fig. 9.2). The simple, related problem that we will analyse is an elastic beam, with flexural rigidity EI, supported on elastic soil which will be treated as a row of independent linear springs – so-called Winkler springs (Fig. 9.11). We will consider a beam under a central point load P and generate the exact solution to the beam equation (9.33). The Winkler spring support implies that at each point on the beam the pressure resisting a settlement y is directly proportional to that settlement through a spring constant (or coefficient of subgrade reaction) λ. For a beam of width B, the force per unit length of beam resulting from the settlement is λBy. Such a foundation lacks the continuity and interaction between adjacent springs which would occur in a real soil foundation – but that would take us away from our single dimension.

We take the origin of the position coordinate x at the centre of the beam (Fig. 9.11). The equation describing the bending of the beam becomes:

$$EI\frac{d^4 y}{dx^4} = -\lambda By \tag{9.34}$$

The negative sign is required because the load generated by the soil springs opposes movement of the beam (compare Fig. 9.10). The problem requires the solution of a fourth order ordinary differential equation with appropriate boundary conditions.

The general solution of (9.34) is:

$$y = \exp[\Omega x](J_1 \cos \Omega x + J_2 \sin \Omega x) + \exp[-\Omega x](J_3 \cos \Omega x + J_4 \sin \Omega x) \tag{9.35}$$

or alternatively:

$$y = (J_5 \cosh \Omega x + J_6 \sinh \Omega x)(J_7 \cos \Omega x + J_8 \sin \Omega x) \tag{9.36}$$

where:

$$\Omega^4 = \frac{\lambda B}{4EI} \tag{9.37}$$

and where J_1, J_2, J_3, J_4 or J_5, J_6, J_7, J_8 must be determined from the boundary conditions of a particular problem. The choice of preferred form of solution, (9.35) or (9.36), depends on the details of the application.

For an infinitely long beam under a central load, the boundary conditions are:

- Slope $dy/dx = 0$ at $x = 0$ from symmetry
- Moment $M = EI d^2 y/dx^2 \to 0$ as $x \to \infty$

9.4 Elastic beam on elastic foundation ♣

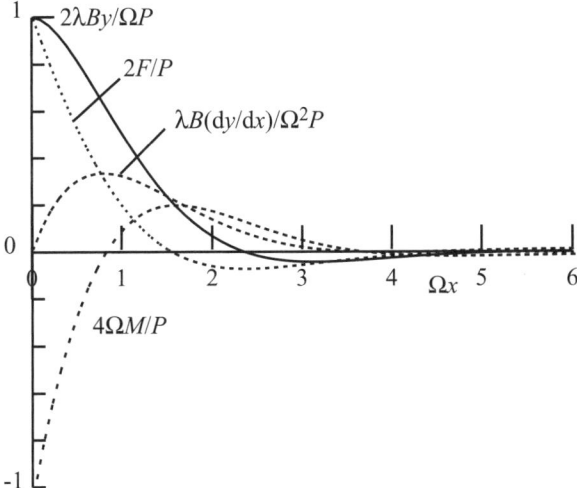

Figure 9.12. Infinite beam on elastic foundation: normalised displacement ($2\lambda By/\Omega P$); slope ($\lambda B(dy/dx)/\Omega^2 P$); moment ($4\Omega M/P$); shear force ($2F/P$).

- Shear force $F = -EI d^3y/dx^3 = -P/2$ at $x = 0$ (recall sign convention in Fig. 9.10)
- Shear force $F = -EI d^3y/dx^3 \to 0$ as $x \to \infty$

The load P is divided equally between the positive x and negative x parts of the beam, and thus the shear force immediately beside the point load is $P/2$. The invocation of considerations of symmetry is very important for the deduction of boundary conditions: the slope of the beam must be zero at the centre, $x = 0$. These boundary conditions give the solution in the half of the beam with $x > 0$; the boundary conditions for $x < 0$ are the same but with the sign of the shear force at the origin reversed, from symmetry.

For this infinitely long beam, the first form of the solution (9.35) is the more convenient. Terms with $\exp[\Omega x]$ must vanish, so that:

$$y = \frac{\Omega P}{2\lambda B} \exp[-\Omega x](\cos \Omega x + \sin \Omega x) \qquad (9.38)$$

The resulting deflections normalised with $\Omega P/2\lambda B$, slopes (normalised with $\Omega^2 P/\lambda B$), bending moment (normalised with $P/4\Omega$) and shear force (normalised with $P/2$) are plotted in Fig. 9.12 as a function of the normalised position on the beam Ωx. The use of this normalised position is an indication that the solution, appropriately presented, is "unique" and applicable to all possible combinations of soil and structural properties. The position along the infinite beam at which equivalent normalised values of settlement, slope, shear force and bending moment occur is itself a function of the relative stiffness of the structure and the soil.

Of course, this is an idealised model: some features of the solution may not be to our liking. For example, the beam lifts up over part of its length – our simple Winkler spring foundation has to be able to take tension as well as compression, whereas

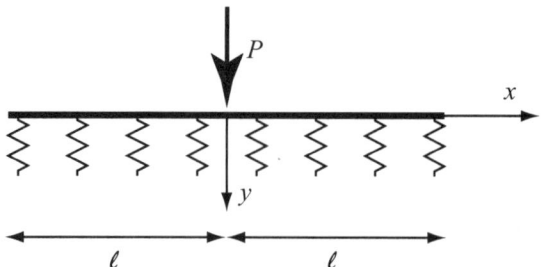

Figure 9.13. Beam of length 2ℓ on elastic foundation.

real soils might have difficulty in resisting such tensile stresses while remaining elastic. Elastic materials can sense the effects of perturbations over great distances (the beam provides continuity even though the foundation springs are independent of each other) and the solution continues to ripple gently even for large values of Ωx. However, we discover that once Ωx is greater than about 6, the moment and shear force remain below 0.5% of their peak values. This might be proposed as the definition of a "long" beam: $\Omega \ell_{long} \approx 6 \Rightarrow \ell_{long} \approx 6(4EI/\lambda B)^{1/4}$. This definition of "long" is a function of both the stiffness properties of the beam (EI) and the stiffness properties of the ground (λB): relative and not absolute values are important.

For a short beam of length 2ℓ (Fig. 9.13), the boundary conditions for $0 < x < \ell$ are:

- Slope $dy/dx = 0$ at $x = 0$ from symmetry
- Moment $M = EI d^2 y/dx^2 = 0$ at $x = \ell$
- Shear force $F = -EI d^3 y/dx^3 = -P/2$ at $x = 0$
- Shear force $F = -EI d^3 y/dx^3 = 0$ at $x = \ell$

and equivalent conditions can be produced for $-\ell < x < 0$. The second form of the solution (9.36) is now the more convenient with the final result:

$$\frac{2\lambda B\ell y}{P} = \Omega \ell \left[(\cosh \Omega x \sin \Omega x - \cos \Omega x \sinh \Omega x) \right. \tag{9.39}$$

$$\left. + (\Gamma_1 \cosh \Omega x \cos \Omega x - \Gamma_2 \sinh \Omega x \sin \Omega x)\right] \tag{9.40}$$

where:

$$\Gamma_1 = \frac{\cosh^2 \Omega \ell + \cos^2 \Omega \ell}{\cosh \Omega \ell \sinh \Omega \ell + \cos \Omega \ell \sin \Omega \ell} \tag{9.41}$$

and:

$$\Gamma_2 = \frac{\sinh^2 \Omega \ell + \sin^2 \Omega \ell}{\cosh \Omega \ell \sinh \Omega \ell + \cos \Omega \ell \sin \Omega \ell} \tag{9.42}$$

The quantity $\tilde{y} = P/2\lambda B\ell$ is the settlement of a load P uniformly distributed over the total length 2ℓ of the beam of width B and resisted by the Winkler springs of stiffness λ; $y/\tilde{y} = 2\lambda B\ell y/P$ thus normalises the profile of settlement of the beam with this reference settlement. If we were to normalise the position coordinate x using a dimensionless variable $\tilde{X} = x/\ell$, then we would see that Ω only enters the

9.4 Elastic beam on elastic foundation ♣

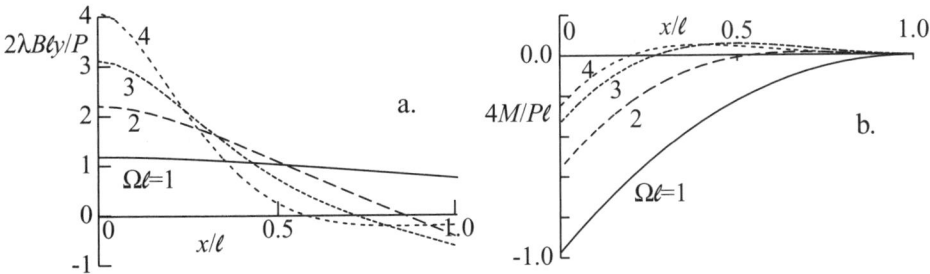

Figure 9.14. Beam on elastic foundation, influence of relative stiffness $\Omega\ell$: (a) normalised displacements and (b) normalised moments.

solution in the dimensionless combination $\Omega\ell$. This (or perhaps more conveniently $\Omega^4\ell^4 = \lambda B\ell^4/4EI$) is the controlling dimensionless scaling parameter for this problem.

Typical distributions of deflection and moment are shown in Fig. 9.14. Low values of $\Omega\ell$ correspond to stiff beams and soft soil – the beam settles almost uniformly for $\Omega\ell = 1$; high values correspond to soft beams and stiff soil – even for $\Omega\ell = 4$, almost all the deformation occurs in the half of the beam nearest the load. For an infinitely stiff beam ($\Omega\ell = 0$), the load P is resisted by a uniform reaction along the length of the beam, and the central moment is then $P\ell/4$. Moments are therefore shown in normalised form as $4M/P\ell$.

As the relative stiffness of structural element and ground changes we move from one extreme, where the beam is infinitely rigid ($\Omega\ell \to 0$, Fig. 9.3a) and settles uniformly but has to be able to support high bending moments, to the other extreme, where the beam is infinitely flexible ($\Omega\ell \to \infty$, Fig. 9.3e) and the moments are negligible (because soil reaction is only generated precisely under the applied load) but the differential settlements between different points on the beam are large. Because the beam is supported on a set of completely independent springs, as the stiffness of the beam tends to zero the differential settlement tends to infinity. The area under each of the curves in Fig. 9.14a must be the same because the total soil reaction must match the applied load. The Winkler spring foundation breaks down as a representation of real soil at this extreme of relative stiffness.

The range of responses of the beam can be presented in terms of the maximum (central) bending moment on the one hand and the differential settlement between the centre of the beam (under the load) and the free end on the other (Fig. 9.15). The central moment, at $x = 0$, is given by:

$$\frac{M_{max}}{P\ell/4} = \frac{1}{\Omega\ell} \frac{\sinh^2 \Omega\ell + \sin^2 \Omega\ell}{\cos \Omega\ell \sin \Omega\ell + \cosh \Omega\ell \sinh \Omega\ell} \quad (9.43)$$

For stiff beams (low values of $\Omega\ell$), the differential settlement is evidently negligible. However, as the beam becomes relatively more flexible the central moment reduces – in fact, it tends to zero – and the differential settlement continues to increase. A key result of soil-structure interaction is this trade-off between structural stiffness (which usually implies cost) and deformations. If we can accommodate differential

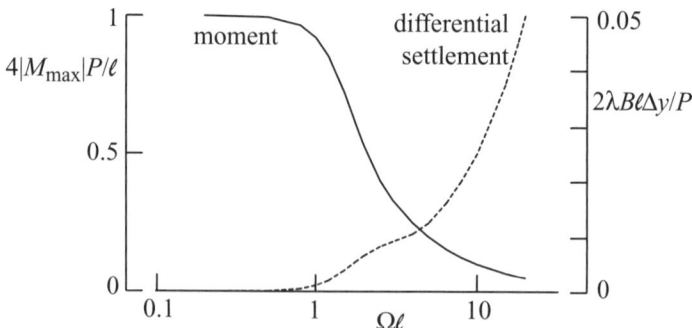

Figure 9.15. Central moment and differential settlement for beam on elastic foundation as function of relative beam stiffness $\Omega\ell$.

displacements of parts of our structure, then we may be able to economise on stiffness.

9.5 Pile under lateral loading ♣

Now we return to the generalised analysis of the tent peg (Fig. 9.4) or laterally loaded pile (Fig. 9.16). A pile of width B in an elastic soil loaded by a lateral force P at the ground surface is analytically very similar to the beam on an elastic foundation. The one dimension of the analysis is now the position z down the pile (Fig. 9.16a). The pressure generated on the pile by the elastic soil is directly proportional to the relative movement of the pile and the soil according to a stiffness λ supplied by a series of independent linear Winkler springs, which act identically whichever way the pile moves relative to the soil (Fig. 9.16b, c). The governing equation for the bending of a beam resisted by loads proportional to the displacement is exactly the same as (9.34), as are the general forms of the solution (9.35), (9.36). The boundary conditions are slightly different.

For an infinitely long pile, the boundary conditions are:

- Moment $M = EI\mathrm{d}^2y/\mathrm{d}z^2 = 0$ at $z = 0$
- Moment $M = EI\mathrm{d}^2y/\mathrm{d}z^2 \to 0$ as $z \to \infty$

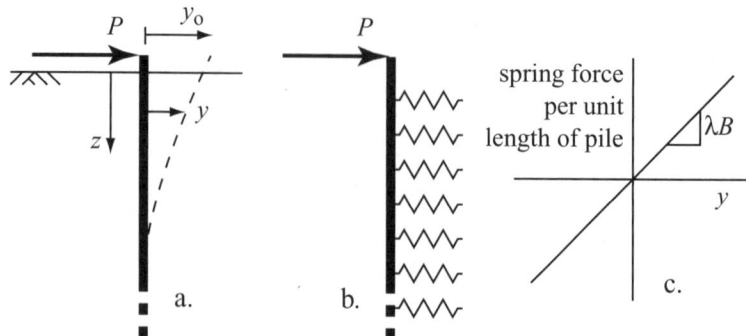

Figure 9.16. Laterally loaded pile in elastic soil.

9.5 Pile under lateral loading ♣

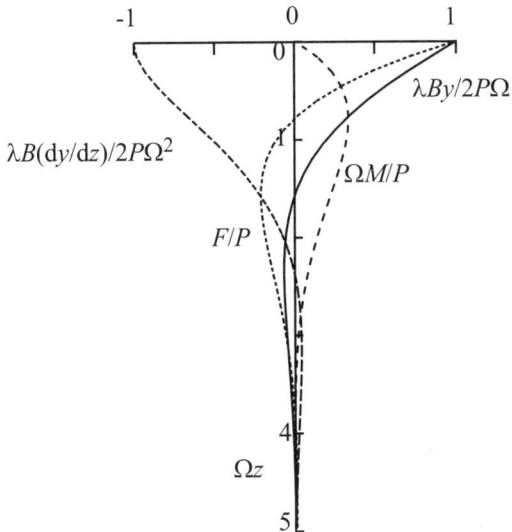

Figure 9.17. Infinite laterally loaded pile in elastic soil: normalised displacement y, slope dy/dz, moment M and shear force F.

- Shear force $F = -EI d^3y/dz^3 = -P$ at $z = 0$
- Shear force $F = -EI d^3y/dz^3 \to 0$ as $z \to \infty$

and the solution is:

$$y = \frac{2P\Omega}{\lambda B} e^{-\Omega z} \cos \Omega z \qquad (9.44)$$

with the controlling parameter $\Omega = (\lambda B/4EI)^{1/4}$ as before.

The variations of deflection (normalised with $2P\Omega/\lambda B$), slope (normalised with $2P\Omega^2/\lambda B$), moment (normalised with P/Ω) and shear force (normalised with P) down the pile are shown in Fig. 9.17. The pile develops an undulating deflected form as the load is transferred down the pile: not much of significance happens below about $\Omega z \approx 6$, just as for the beam (Fig. 9.12). This would constitute our definition of a "long" pile, and as we expect, the definition depends on the stiffness properties of both soil and pile:

$$\ell_{\text{long}} \approx 6 \left(\frac{4EI}{\lambda B} \right)^{1/4} \qquad (9.45)$$

For a short pile of length ℓ, there can be no residual shear force or bending moment at the deep tip of the pile, and the boundary conditions are:

- Moment $M = EI d^2y/dz^2 = 0$ at $z = 0$
- Moment $M = EI d^2y/dz^2 = 0$ at $z = \ell$
- Shear force $F = -EI d^3y/dz^3 = -P$ at $z = 0$
- Shear force $F = -EI d^3y/dz^3 = 0$ at $z = \ell$

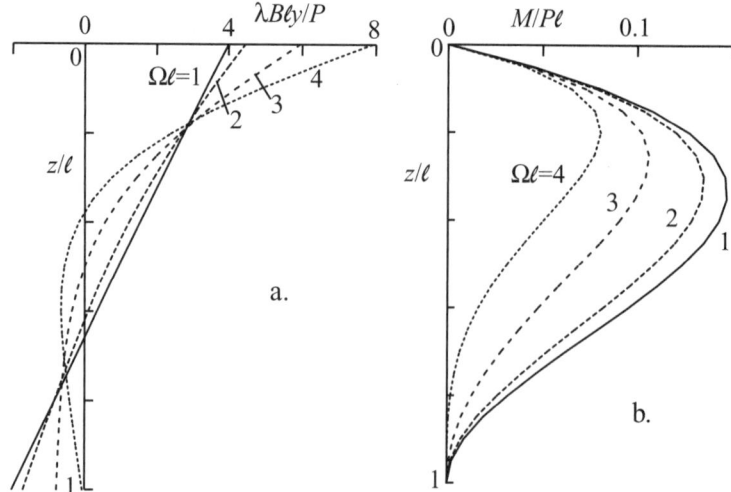

Figure 9.18. Laterally loaded pile of length ℓ in elastic soil showing influence of relative pile stiffness $\Omega\ell$: (a) normalised displacements and (b) normalised moments.

The solution is:

$$\frac{\lambda B \ell y}{2P} = \frac{\Omega\ell}{\sin^2 \Omega\ell - \sinh^2 \Omega\ell} \times$$

$$\left[\sinh^2 \Omega\ell \sinh \Omega z \cos \Omega z + \sin^2 \Omega\ell \cosh \Omega z \sin \Omega z + \right.$$

$$\left. (\cos \Omega\ell \sin \Omega\ell - \cosh \Omega\ell \sinh \Omega\ell) \cosh \Omega z \cos \Omega z\right] \quad (9.46)$$

We deduce that the dimensionless parameter that controls the behaviour of the pile is $\Omega\ell$ or $\Omega^4 \ell^4 = \lambda B \ell^4 / 4EI$, which is clearly a function of relative stiffness of pile and soil. Typical normalised displacements and moments for values of $\Omega\ell$ between 1 and 4 are shown in Fig. 9.18: for a stiff pile (low values of $\Omega\ell$), the pile hardly bends at all but kicks backwards to generate the moment to resist the applied load (compare the stiff tent peg in Fig. 9.4b). For more flexible piles (higher $\Omega\ell$), the lateral deflection of the top of the pile increases and the flexure of the pile also increases (compare the flexible tent peg in Fig. 9.4c).

The pile stiffness P/y_o can be found by setting $z = 0$ in (9.46):

$$\frac{P}{y_o} = \frac{\lambda B}{2\Omega} \frac{\sin^2 \Omega\ell - \sinh^2 \Omega\ell}{\sin \Omega\ell \cos \Omega\ell - \sinh \Omega\ell \cosh \Omega\ell} \quad (9.47)$$

9.6 Soil-structure interaction: next steps

In each of the generic problems that we have analysed, we have described the soil in its interaction with the structural elements by means of a series of independent linear elastic springs. This is obviously a travesty of the real problem but is pedagogically valuable in showing clearly how the relative stiffness properties of both the structural elements and the soil springs interact to control the overall character of the soil-structure interaction. The analyses also show the way in which changing the relative stiffness exchanges deformations for structural resultants. Control of displacements requires stiffer structures able to sustain higher forces or moments: the designer needs to decide on the desirable balance.

We have described both the structural elements and the soil as strictly linear and elastic in their response to changing loads. Elasticity formally requires that the application and removal of any load should leave no permanent deformation. Linearity of response is an extra constraint. In reality, although the structural elements may behave elastically, it is very unlikely that the behaviour of the soil interacting with the structural element will be linear and elastic except under extremely small deformations: we have seen the nonlinearity of soil stiffness in Sections 4.5 and 8.11. It is possible to extend our analysis to include non-linearity of the spring model while still maintaining the one-dimensional treatment, but this would require the use of numerical solution techniques. However, no matter by what route nonlinear or two/three-dimensional problems of soil-structure interaction are analysed, it will always be found that the emergent response is dependent on the relative stiffness properties of the soil and the structural elements.

9.7 Summary

Here is a concise list of the key messages from this chapter, which are also encapsulated in the mind map (Fig. 9.19).

1. Many geotechnical prototypes include structural elements. Calculation of the response of such prototypes under working loads – *serviceability limit states* – requires consideration of stiffness characteristics of both the geotechnical and structural materials.
2. The significance of soil-structure interaction depends on relative stiffness of structure and ground.
3. Analyses of elastic structures interacting with elastic ground are capable of exact solution in terms of dimensionless relative stiffnesses.
4. Examples of soil-structure interaction provide opportunities to integrate aspects of civil engineering education, which are often treated entirely independently.

9.8 Exercises: Soil-structure interaction

1. A axially loaded pile of radius $r_o = 0.5$ m and length $\ell = 40$ m is made of concrete with Young's modulus $E_p = 25$ GPa. It is driven through soil with shaft shear transfer stiffness $\lambda = 4$ MPa and ends with its toe in firmer ground with base transfer stiffness $\lambda_b = 100/\pi$ MPa. The base of the pile is under-reamed to a radius of $r_b = 1$ m. What is the axial stiffness of the pile P_t/ς_t, and what proportion of the load reaches the toe of the pile?

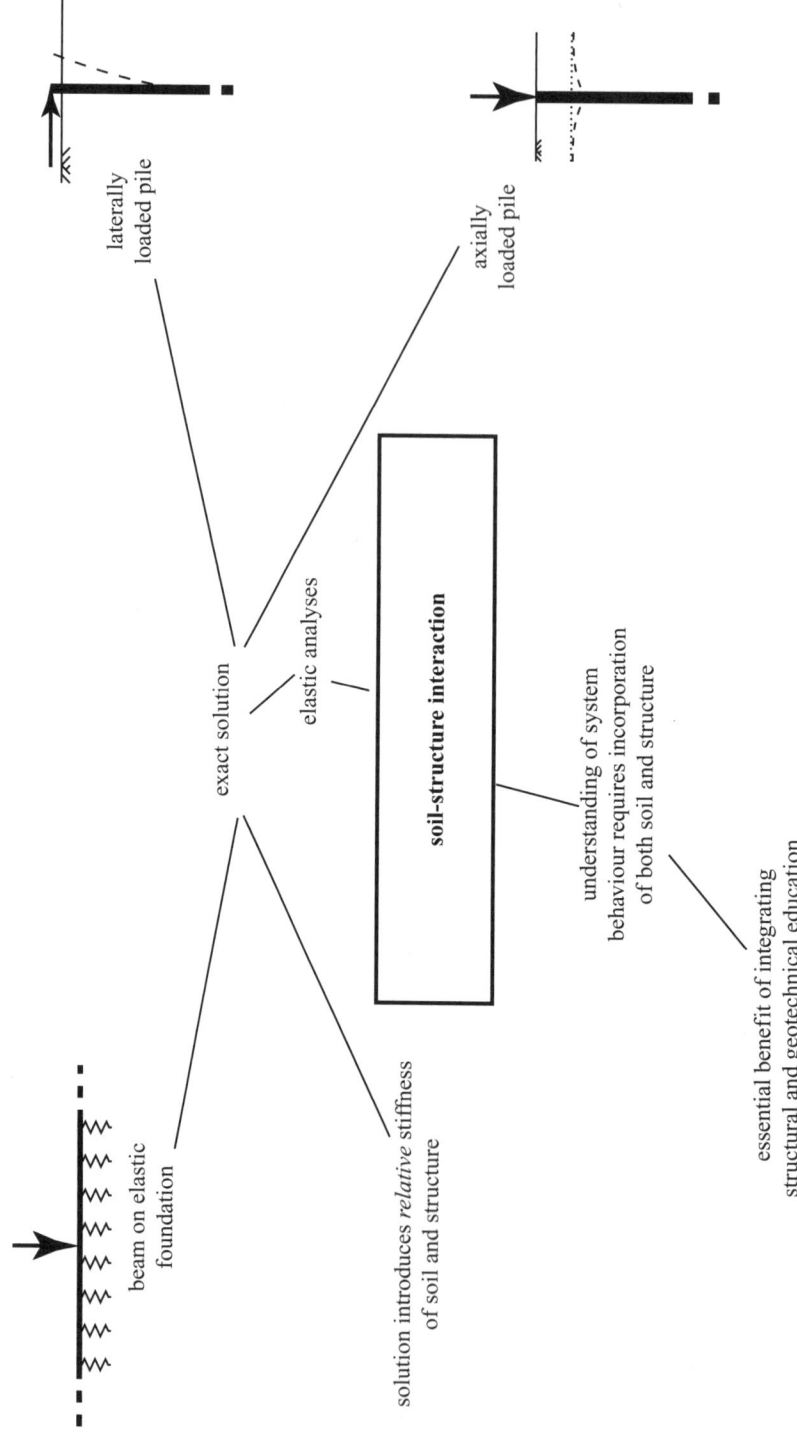

Figure 9.19. Mind map: soil-structure interaction.

9.8 Exercises: Soil-structure interaction

2. Use the analysis of the beam on an elastic foundation to find the range of values of normalised lengths $\Omega\ell$ for which the soil springs are in compression beneath the whole length of the beam.
3. A stiff laterally loaded pile will "kick back" at its toe as it rotates under the applied load (for example, $\Omega\ell = 1$ in Fig. 9.18a). For a pile head stiffness $P/y_o = 500$ MN/m and pile-soil stiffness $\lambda = 750$ MPa/m, what length of pile of width $B = 0.5$ m is required to keep the kick-back at the toe of the pile to less than 1% of the top displacement, $|y_{z=\ell}| < y_o/100$?

10 Envoi

10.1 Summary

Each chapter has ended with a summary of the key points. Here it is appropriate to reiterate the primary intentions of this book and to point the way towards the next stages of soil mechanics education.

There are two themes which have been developed in this book:

- The model of soil stiffness and strength links *effective stresses* and *density*. Changes in mechanical behaviour of soils can only be properly understood when the stresses and volumetric packing are considered in parallel. The role that history plays in changes in density – which tend to be locked into soils as loads are applied and then removed – is key for the subsequent understanding of the strength of soils. There is then a link with permeability and the ease with which water (or other pore fluid) can move through the voids of the soil, and thus the ability of the soil to respond rapidly to changes in stress. The concepts of *drained* and *undrained* response again bring together the thoughts about effective stresses and volumetric packing. This is the essence of *critical state soil mechanics*,[1] which at its simplest level (Fig. 10.1) tells us that we cannot hope to understand the behaviour of soils unless we think all the time in parallel of the changes in effective stress and of density.

- The second theme is that it is perhaps surprising how far we can get in the presentation of fundamental ideas of soil mechanics without straying beyond the single dimension. We have seen that there are a number of modestly realistic applications which can be discussed with only one dimension in the governing equations – examples of consolidation and gain in strength, slope stability, and soil-structure interaction show basic features of response on which we may wish to elaborate in future multi-dimensional analyses but whose messages will remain intact. The final application, soil-structure interaction, is seen as particularly important because the teaching of civil engineering in universities tends too much to separate soil mechanics/geotechnical engineering and structural engineering in a way that leads specialists in each area to lose sight of the need

---

[1] Schofield, A.N. & Wroth, C.P. (1968) *Critical state soil mechanics*. McGraw-Hill, London; Muir Wood, D. (1990) *Soil behaviour and critical state soil mechanics*. Cambridge University Press.

10.2 Beyond the single dimension

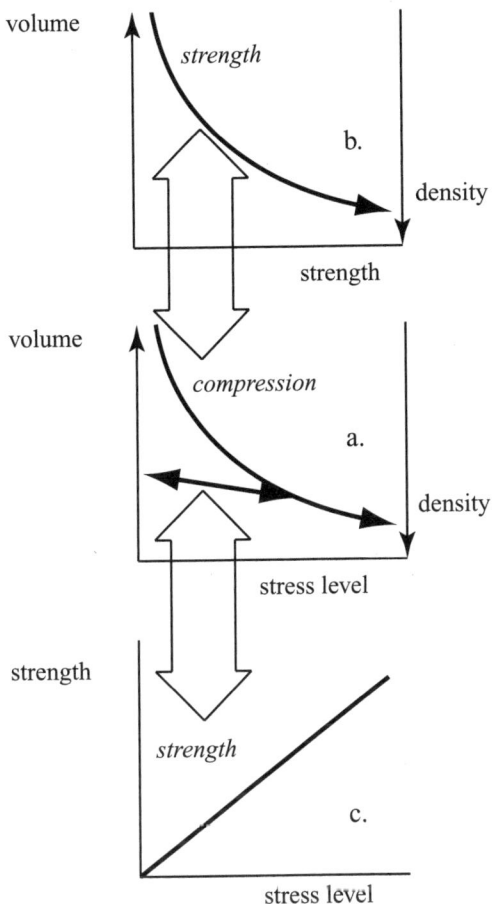

Figure 10.1. Critical state soil mechanics: (a) density and compression; (b) density and strength; (c) stresses and strength.

for integration, recognising that it is the combined response of geotechnical and structural elements that controls the overall behaviour of the engineered system. The need to consider relative stiffnesses of soil and structural elements is essential.

10.2 Beyond the single dimension

By the end of the book, the applications are obviously pushing at the one-dimensional limits that have been imposed. The two important aspects of soil behaviour and geotechnical systems that require more than a one-dimensional approach are:
- the analysis of stress in continuous materials; and
- the actual two- or three-dimensional geometry of real problems.

The restriction of discussion of settlement to one-dimensional configurations is extremely artificial but serves the purpose of introducing key ideas which are just as relevant for more realistic situations.

Exercises: numerical answers

Chapter 2

Answers: Section 2.5

1. 3.92 kN
2. 15.7 kPa on two feet, 31.4 kPa on one foot
3. 7.34 Mg
4. 2.43 kPa
5. 0.45 m on one foot, 0.32 m on two feet
6. 173×10^{-6} N

Answers: Section 2.10

1. 0, 58.9, 247.2, 360.0 kPa
2. σ_z: 0, 58.9, 247.2, 360.0 kPa; u: 0, 29.4, 147.2, 196.2 kPa; σ'_z: 0, 29.4, 100.1, 163.8 kPa
3. σ_z: 0, 58.9, 247.2, 360.0 kPa; u: 0/−14.7, 14.7, 132.4, 181.5 kPa; σ'_z: 0/14.7, 47.1, 114.8, 178.5 kPa
4. σ_z: 53.0, 131.5, 168.7 kPa; u: 0, 39.2, 58.9 kPa; σ'_z: 53.0, 92.2, 109.9 kPa
5. σ_z: 57.4, 135.9, 173.1 kPa; u: 14.7, 53.9, 73.6 kPa; σ'_z: 42.7, 81.9, 99.6 kPa; $\Delta\sigma'_z$: −10.3, −10.3, −10.3 kPa
6. maximum 66.7 kPa; minimum 43.2 kPa

Chapter 3

Answers: Section 3.8.1

1. (c); 2. (a); 3. (c); 4. (b); 5. (c); 6. (b); 7. (c); 8. (a); 9. (b), (c), (a), (b), (c), (a), (b); 10. (a), (c), (a), (a), (b), (a), (b); 11. (b), (a), (c), (a), (a), (b), (b), (b), (b)

Answers: Section 3.8.2

1. $w = 0.2$, $v = 1.8$, $e = 0.8$, $S_r = 0.675$
2. $e = 0.93$, $v = 1.93$, $\rho_d = 1.37$ Mg/m^3, $\rho = 1.78$ Mg/m^3

Exercises: numerical answers

3. $\rho = 2.11$ Mg/m^3, $\rho_d = 1.89$ Mg/m^3, $v = 1.43$, $e = 0.43$, $w = 0.11$, $S_r = 0.72$
4. $w = 31.3\%$, $e = 0.837$, $v = 1.837$
5. $e = 1.078$, $v = 2.078$, $w = 39.9\%$, $\rho = 1.82$ Mg/m^3, $\rho_d = 1.30$ Mg/m^3
6. $S_r = 96\%$, $e = 0.847$, $v = 1.847$, $\rho_d = 1.46$ Mg/m^3
7. $\rho = 2$ Mg/m^3, $e = 0.594$, $v = 1.594$, $S_r = 77\%$, $\rho_d = 1.71$ Mg/m^3
8. 2.55 km, 4658 loads

Chapter 4

Answers: Section 4.9

1. 10 GPa, 108 MPa, 12.2 MPa, 1.5 MPa
2. 0.00015%, 0.014%, 0.123%, 1%
3. 0.0015%, 0.13%, 1.0%, 6.9%
4. 0.345 m
5. 1.55%, 0.062 m
6. (a) before 10.8 kPa, 27.5 kPa, 39.2 kPa, 51.0 kPa, 62.8 kPa; after 17.6 kPa, 41.2 kPa, 53.0 kPa, 64.7 kPa, 76.5 kPa; (b) 0.15%, 0.39%, 0.32%, 0.27%, 0.24%; 0.027 m
7. 62.8 kPa, 45.1 kPa; ocr varies between 1 and 1.4; $\alpha \approx 0.6?$, $\chi \approx 100?$; E_o varies between about 6200 kPa and 7560 kPa, average $\approx$ 6900 kPa; settlement 20 mm
8. heave on excavation 0.013 m; settlement for 25 kPa 0.033 m; settlement for 40 kPa 0.070 m
9. $\chi_{nc} = 9.15$, $\chi_{oc} = 33.94$, $\alpha_{nc} = \alpha_{oc} = 1.05$ (or $\alpha_{nc} = \alpha_{oc} = 1$, $\chi_{nc} = 9.2$, $\chi_{oc} = 34.5$)

Chapter 5

Answers: Section 5.12

1. 1.96, 4.91, 7.85 kPa; 1.96, 0.98, 0 kPa; down; 0.067 m/s
2. 5×10^{-8} m/s downward
3. 98.1, 73.6 kPa; 10^{-7} m/s downwards
4. 14.71 kPa, no flow; 24.52 kPa, no flow; 39.24 kPa, up 0.6×10^{-4} m/s; 53.95 kPa, no flow; 34.34 kPa, down 0.8×10^{-4} m/s
5. 4 m, 11 m, 39.24 kPa, 107.91 kPa, 7.5×10^{-8} m/s
6. total heads: 1.5 m, 1.5 m; pressure heads: 5.5 m, 9.5 m; flow rate: 3.75×10^{-8} m/s; total head drop across sand: 0.015 m
7. horizontal: $k_h = 1.35 \times 10^{-3}$ m/s; vertical: $k_v = 14.3 \times 10^{-5}$ m/s
8. total head: -1 m, $+3$ m; pressure head: 2 m, 10 m; pore pressure: 19.6 kPa, 98.1 kPa; total stress: 53.46 kPa, 131.94 kPa; effective stress: 33.84 kPa, 33.84 kPa; flow rate 1×10^{-8} m/s
9. $k = 4.8 \times 10^{-8}$ m/s

Chapter 6

Answers: Section 6.5

1. 140 kPa
2. 60 kPa, 140 kPa, 0%
3. 200 kPa, 10 kPa, 190 kPa, 3.33%
4. 152.8 kPa, 152.8 kPa
5. 229.2 kPa, 152.8 kPa, 76.4 kPa, no change
6. 229.2 kPa, 229.2 kPa, 0 kPa, 4.4%, 19.13 mm
7. values of σ_z, u, σ'_z at the base of the sand and at the top and bottom of clay layer
   initially: 33, 0, 33; 33, 0, 33; 213, 98.1, 114.9 kPa
   immediately after sand is removed: 16.5, 0, 16.5; 16.5, −16.5, 33; 196.5, 81.6, 114.9 kPa
   long term: 16.5, 0, 16.5; 16.5, 0, 16.5, 196.5, 98.1, 98.4 kPa
   heave: no change immediately, 0.097 m in long term
8. values of σ_z, u, σ'_z at the top and bottom of the clay in the long term 33, 0, 33; 213, 65, 148 kPa; long term settlement 0.097 m
9. values of σ_z, u, σ'_z at base of sand and at top and bottom of clay
   initially: 57.01, 19.62, 37.39; 57.01, 19.62, 37.39; 137.01, 58.86, 78.15 kPa
   immediately after excavation: 48.73, 19.62, 29.11; 48.73, 11.34, 37.39; 128.73, 50.58, 78.15 kPa
   long term: 48.73, 19.62, 29.11; 48.73, 19.62, 29.11; 128.73, 58.86, 69.87 kPa
   long term heave: 0.021 m
10. initially $\sigma_z = 19z, u = 9z, \sigma'_z = 10z$; finally $\sigma_z = 19z, u = 9.81z, \sigma'_z = 9.19z$; average increase in pore pressure and reduction in vertical effective stress 121.5 kPa; heave 0.3375 m

Chapter 7

Answers: Section 7.7

1. $c_v = 1.5 \times 10^{-7}$ m²/s; consolidation front reaches centre at $\tilde{T} = 1/12, t = 54.5$ s; 80% consolidation for $\tilde{T} = 0.485, t = 317$ s
2. $L = 1.5$ m, time for 80% consolidation $t = 82.5$ days; independent of stress increment
3. $c_v = 3.14 \times 10^{-9}$ m²/s; 10.4 years
4. 139 days

Exercises: numerical answers

5. Stage 1: $\tilde{S} = \sqrt{(\pi-2)\tilde{T}}$ until $\tilde{T} = \tilde{T}_A = (\pi-2)/\pi^2$; Stage 2: $\tilde{S} = 1 - \frac{2}{\pi}\exp\{-\pi^2(\tilde{T}-\tilde{T}_A)/4\}$
6. Intersection slope ×1.1: $\tilde{S} = 0.8527$, $\tilde{T} = 0.6910$, $c_v = 16.5 \times 10^{-9}$ m²/s, $k = 51.2 \times 10^{-12}$ m/s; intersection slope ×1.2: $\tilde{S} = 0.9259$, $\tilde{T} = 0.9695$, $c_v = 16.2 \times 10^{-9}$ m²/s, $k = 50.8 \times 10^{-12}$ m/s; $\tilde{S} = 0.9$ for $\tilde{T} = 0.848$, 10.2 yrs (slope×1.1), 10.4 yrs (slope×1/2).
7. Intersection slope ×1.1: $\tilde{S} = 0.8965$, $\tilde{T} = 0.852$, $c_v = 20.4 \times 10^{-9}$ m²/s, $k = 60.2 \times 10^{-12}$ m/s; intersection slope ×1.2: $\tilde{S} = 0.9485$, $\tilde{T} = 1.135$, $c_v = 19.0 \times 10^{-9}$ m²/s, $k = 58.1 \times 10^{-12}$ m/s; $\tilde{S} = 0.9$ for $\tilde{T} = 0.866$, 8.4 yrs (slope×1.1), 9.0 yrs (slope×1.2).

Chapter 8

Answers: Section 8.13

1. 120 kPa, 36 kPa
2. 130 kPa, 120 kPa, 36 kPa
3. 250 kPa, 75 kPa; density increases as water is squeezed out as effective stress and strength increase
4. mobilised strength 18.3 kPa; mobilised angle of friction 19.7° – close to available frictional strength 21°
5. $r_u = -0.22$ throughout; pore pressure at depth 5 m: $u = -20.4$ kPa; with seepage parallel to slope $\phi_{mob} = 58.8°$; $r_u = 0.327$; support force 2.94 MN/m
6. $d_s = d/\sqrt{3}$
7. normal compression line

$$\ln\left(\frac{v_1}{v}\right) = \frac{1}{\chi(1-\alpha)}\left[\left(\frac{\sigma'_z}{\sigma_{ref}}\right)^{1-\alpha} - \left(\frac{\sigma'_{z1}}{\sigma_{ref}}\right)^{1-\alpha}\right]$$

if $\alpha = 1$, $v = v_1[\sigma'_{z1}/\sigma'_z]^{(1/\chi)}$, where $v = v_1$ for $\sigma'_z = \sigma_{z1}$.
(a) normal compression $v = 1.866$; drained $\tau_f = 46.63$ kPa, $v_f = 1.752$; undrained $\tau_f = 23.27$ kPa, $\sigma'_{zf} = 49.90$ kPa. (b) normal compression $v = 1.820$; drained $\tau_f = 233.15$ kPa, $v_f = 1.497$; undrained $\tau_f = 30.76$ kPa, $\sigma'_{zf} = 65.96$ kPa. (c) overconsolidated $v = 1.832$; drained $\tau_f = 46.63$ kPa, $v_f = 1.752$; undrained $\tau_f = 28.71$ kPa, $\sigma'_{zf} = 61.57$ kPa.
(d) Normal compression to 100 kPa: drained $\tau = 23.53$ kPa, $\sigma'_z = 100$ kPa, $v = 1.863$, $\varepsilon_z = 0.19\%$, $\varepsilon_s = 0.59\%$; undrained $\tau = 18.24$ kPa, $\sigma'_z = 78.25$ kPa, $v = 1.866$, $\varepsilon_z = 0$, $\varepsilon_s = 0.71\%$.
Normal compression to 500 kPa: drained $\tau = 117.0$ kPa, $\sigma'_z = 500$ kPa, $v = 1.814$, $\varepsilon_z = 0.37\%$, $\varepsilon_s = 1.21\%$; undrained $\tau = 54.57$ kPa, $\sigma'_z = 234.1$ kPa, $v = 1.820$, $\varepsilon_z = 0$, $\varepsilon_s = 4.54\%$.
Overconsolidation to 100 kPa: drained $\tau = 23.52$ kPa, $\sigma'_z = 100$ kPa, $v = 1.828$, $\varepsilon_z = 0.18\%$, $\varepsilon_s = 0.56\%$; undrained $\tau = 11.76$ kPa, $\sigma'_z = 50.55$ kPa, $v = 1.832$, $\varepsilon_z = 0$, $\varepsilon_s = 0.42\%$.

Chapter 9

Answers: Section 9.8

1. pile stiffness $P_t/\varsigma_t = 644$ MN/m; proportion of load at toe $P_b/P_t = 0.062$
2. $\Omega \ell \leq \pi/2$
3. $\Omega \ell = 3.772$, $\ell = 10.07$ m

Index

acceleration, 91
angle of repose, 191–193
aquifer, 111
Archimedes, 24, 26, 59

Balodis, A., 42
beam, 216–219
bending moment, 222, 223
Bernoulli's equation, 90, 93–95, 97, 101
body force, 13
Bolton, M.D., 76
boulders, 53, 54
boundary conditions, 162, 220, 222, 224, 225
bridge
    abutment, 6
    arch, 5
buoyancy, 26, 59, 188
Buzan, T., 11

capillary rise, 31
Carman, 102
Carrier, W.D., 102
Carslaw, H.S., 159
clay, 47, 49, 54–56, 64, 75, 83, 99, 101, 102, 123, 127, 140, 141, 145, 177, 182, 187, 193
    Drammen, 42
    mineral, 43, 54, 102, 195
    Weald, 42
coefficient
    of consolidation, 144, 145, 147, 149, 150, 157
    of permeability, *see* permeability
    of sub-grade reaction, 220
    of uniformity, 54, 55
collapse, 169
Collin, 170–172
compaction, 49–52, 64
compression
    confined, 72, 73, 76
    unconfined, 71, 72
conic section, 143
consolidation, ix, 127, 138
    degree of, 147, 163
    finite layer, 161
    front, 141, 144, 147, 160, 161, 165
    secondary, 151
Coulomb, 3
creep, 151
critical state soil mechanics, ix, 230

dam, 2, 9, 10, 49, 101, 110
Darcy's Law, 100–105, 123, 140, 141, 151, 156, 160, 165
datum, 90, 94, 123
deformation, 4, 6
density, 47, 87, 230
    bulk, 44, 45, 49
    buoyant, 188, 190
    dry, 45, 47, 50, 51, 174
    mineral, 174
    relative, 62
    saturated, 45, 47, 190
    water, 24, 41
diffusion equation, 138, 165
dilatancy, 174, 175, 177, 198–200
dimensional analysis, 57, 58, 194
down-drag, 182–185
drained response, 129, 136, 177–179, 200, 201, 205, 230
Dupuit, J., 103

Earth, 14, 15, 37, 41, 52
earthquake, 7, 8, 129
effective stress, 28–32, 37, 46, 76, 90, 128, 134, 141, 142, 165, 174, 177, 202, 230
    Principle of, 29, 37, 46, 129, 134, 140, 157, 193
elasticity, 67, 68, 74, 76, 82, 83, 87, 182, 208, 211, 227
electrical resistors, 108
Elson, W.K., 212
embankment, 6
end bearing, 169, 211
equilibrium, 12, 13, 16, 18–20, 26, 28–31, 37, 46

erosion, 52
error function, 159
excavation, 4, 8

fall-cone, 193–195
Fleming, W.G.K., 212
Forchheimer, P., 103
foundation, 6, 7, 170, 171, 208–210, 220–224
Fourier series, 163, 165
friction, 174, 203
   angle of, 174
   mobilised, 187, 189–191

gravel, 47, 52, 54, 64, 99, 101, 102, 123, 127, 129, 177
gravitational constant, 14
gravity, 13–16, 41, 93, 194

hardness, 52, 54
Hazen, 102
head
   elevation, 93–95, 101, 109, 123, 189
   pressure, 93–95, 101, 123, 189
   total, 93–95, 97, 100, 101, 104, 106–113, 123, 138, 140, 141, 156, 165, 188, 189
Henkel, D.J., 205
Hooke's Law, 72, 74, 75, 82, 87
hydraulic gradient, 100, 104, 109, 123, 141, 146, 156
   critical, 110
hydrostatics, 23–26, 90

isochrone, 142, 145, 148, 163, 165
   parabolic, 142, 144, 146, 149, 151, 156, 160, 161, 165

Jaeger, J.C., 159
Janbu, N., 76, 77

kinetic energy, 93, 94, 101
Kolbuszewski, J.J., 62
Kozeny, 102

laboratory test, 47, 54, 56, 57, 61, 149, 151, 191, 192
landslide, 90, 91, 170
liquid limit, 195

mass, 15, 37
McDowell, G.R., 76
mind map, 11, 37, 62, 87, 121, 134, 165, 203, 227
minerals, 52
modelling, 126
modulus number, 77, 83, 87
Moh hardness scale, 52
moisture content, *see* water content
Moon, 15, 41
Muir Wood, D., 230

Newton
   Law of Gravitational attraction, 13, 14
   Law of Motion, 12, 15, 16, 37, 90
normal compression, 82, 83, 179, 180, 196

oedometer, 71, 72, 87, 127, 128, 149–151, 179
overconsolidation, 82, 83, 87, 179–181, 196
   ratio, 83, 87, 180, 196

Palmer, A.C., 57, 194
parabola, 143, 144
particle
   breakage, 76
   shape, 41, 52, 61, 62, 127
   size, 52, 54, 55, 62, 64, 99, 123, 127
      distribution, 54–56, 59, 64
penetration test, 193
permeability, 100, 101, 123, 127–129, 134, 136, 141, 151, 158, 176, 179, 230
   layered soil, 107, 123
   specific, 100
permeameter
   constant head, 105, 106, 123
   falling head, 104, 106, 123
piezometer, 94, 95
pile, 6, 8, 169, 170, 181–185, 210–215, 224–226
   axial load, 211–215
   sheet, 8, 9
piping, 110, 123
Poiseuille's equation, 96, 99–102, 111, 128
Poisson's ratio, 69, 71–75, 82, 182
pore pressure, 29, 37, 46, 90, 127, 128, 134, 165
   excess, 128, 138–142, 156, 157, 165
porosity, 44, 64
potential energy, 93
preconsolidation pressure, 180
pycnometer, 48

quartz, 52

radial flow, 111, 123
Randolph, M.F., 212
Ransome, A., 16, 17
retaining structure, 3, 7, 49
Reynolds, 176
Reynolds' number, 103, 104
Roberts, A.J., ix
Robertson, D., 76

Saint Exupéry, A. de, 16
sand, 42, 47, 49, 52, 54, 56, 64, 75, 102, 103, 127, 129, 140, 145, 177, 182
   dense, 61, 175
   loose, 61, 175
saturation
   degree of, 44, 48, 50, 51, 64
scanning electron micrograph, 42, 55
Schofield, A.N., 142, 230
sedimentation, 60, 64

Index

seepage, 9, 189–191
   force, 109, 123
   velocity, 101
serviceability limit state, 3, 67, 169, 183
settlement, 142, 145, 160
   differential, 209, 223
   dimensionless, 149, 160
shaft resistance, 169, 181–183, 211
shape function, 142
shear box, 171–173, 175, 195, 199, 203
shear force, 222
shearing resistance
   angle of, 174, 182, 187, 191
shingle, 42, 56, 103
sieving, 57, 64
silt, 54
simple harmonic motion, 162
Skempton, A.W., 205
slip surface, 170, 171, 186, 203
slope, 91, 170, 171, 230
   infinite, 185–191
soil, 1
soil mineral, 43, 77, 127, 191
soil-structure interaction, ix, 8, 208–227, 230
   pile under lateral loading, 226
specific gravity, 41, 43, 47, 48, 191
   bottle, *see* pycnometer
specific surface, 56, 102
specific volume, 43, 45, 50, 51, 64, 174, 178, 180, 197
spherical flow, 112
standpipe, 24, 28, 93–95
Statens Järnvägars Geotekniska Kommission, 193, 194
steel balls, 176, 177
stiffness, 3, 67, 169, 230
   confined, 72, 74–76, 127, 142, 143, 151, 158, 196
   relative, 208, 213, 215, 222, 226, 227, 231
stiffness exponent, 76
Stokes' Law, 57, 59
strain
   axial, 68
   natural, 70, 71
   true, 70, 71, 78, 79
   volumetric, 69
strength, 3, 13, 67, 169, 174, 230
   frictional, 174
   mobilised, 198
   peak, 173, 197
   ultimate, 173, 174, 197

stress, 16–18
   effective, *see* effective stress
   normal, 18
   principal, 73
   shear, 18, 20, 57
   total, *see* total stress
stress-dilatancy relationship, 198, 199
surface tension, 31, 32, 192, 193
symmetry, 20, 145, 186, 221, 222

Taylor, D.W., 172, 173
time
   dimensionless, 147–149, 158, 161
total stress, 28–32, 37, 46, 90, 128, 134, 165
transport, 52, 64
   by air, 52
   by gravity, 52
   by ice, 52–54
   by water, 52–54
tunnel, 4, 5

ultimate limit state, 67, 169, 182, 183, 185
under-reaming, 214
undrained response, 129, 136, 177–181, 187, 193–195, 200–202, 205, 230
unit weight, 41, 47
   bulk, 46
   buoyant, 46
   water, 41, 46

viscosity, 57, 58, 96, 101, 103
   of water, 57, 58
void ratio, 43, 47, 48, 50, 56, 64, 101, 191
   air, 44, 50, 51
   maximum, 62
   minimum, 62
volume fraction, 44

water
   content, 44, 47, 50, 51, 64
   sea, 24, 41
   table, 23, 28–30, 32, 46, 90
weathering, 52, 56, 64
weight, 15, 37
Weltman, A.J., 212
Winkler springs, 220–223
Wood, D.M., 195
Wroth, C.P., 142, 176, 177, 230

Young's modulus, 68, 71–75, 82, 213

Zélikson, A., 126